"I believe that Christians are called to make the world a better place. Engineers and designers have a unique opportunity to pursue that calling. *A Christian Field Guide to Technology for Engineers and Designers* provides a helpful framework to experience appropriate pride in pursuing the meaningful field of engineering and design while also acknowledging the humility associated with human limitations."

Jason Andringa, president and CEO, Vermeer Corporation

"A beautifully and clearly written book on technology! After a long history and a recent spurt in development, the wonders of technology have left an all-controlling mark on our culture, so that we speak of the technological society. Thinking about that culture fills us with awe as well as anxiety. Technology knows the waywardness of Babel's tower as well as the blessing of salvation through Noah's ark. Responsible technology is rightly given much attention in this book: it is a technology in harmony with nature, in the service of all that lives, and above all, subordinate to our neighbor and in praise of the Creator. This book makes it clear that it is not enough for an engineer to be *just* an engineer."

Egbert Schuurman, emeritus professor in Christian philosophy at the Technological Universities of Delft and Eindhoven in the Netherlands, former senator of the Dutch Parliament

"Here is a rich resource for the early-career engineer who needs help applying biblical truth to their technical endeavors in the workplace. Through an excellent overview of important theological concepts, thoughtful design principles, and practical vocational wisdom, Brue, Schuurman, and VanderLeest explain how integrating faith and work in practicing engineering and creating technology provides endless opportunities to both glorify God and enjoy him. If only such a book had been around when I entered the workforce with an engineering degree!"

Steven Lindsey, executive director of the Center for Faith+Work Los Angeles, retired Boeing systems engineering executive

"I have taught engineering ethics at the college level for more than thirty years. This is the book that I wish had been available when I started. Many modern engineering ethics books tend to ignore the relevance of Christian faith to the development and use of modern technology. This excellent book will be a helpful addition for Christian faculty and students at public universities, so they can see how their faith can relate to the use of technology. Similarly, it can encourage practicing Christian engineers who wonder how their work can be informed by their faith. For Christian universities this book would make an excellent text for either a freshman introduction to engineering course or a stand-alone engineering ethics text. If I were still teaching full-time, I would use this book in my engineering ethics class."

William Jordan, professor emeritus of mechanical engineering, Baylor University

D0838967

"We have been blessed in recent years with some wonderful explorations of what it means to glorify God in our daily work. Now this marvelous book shows us how to take further steps by focusing on the details of a specific area of work. The authors not only give us profound—and often very inspiring—guidelines for believers engaged in engineering and design, they also point the way for folks in other professions to dig into the specifics of their vocations."

Richard J. Mouw, president emeritus of Fuller Theological Seminary

"So often, Christians focus on the ethical and social implications of technology from a user's perspective but fail to recognize the faithful service of brothers and sisters who work tirelessly behind the scenes to design and develop these powerful culture-shaping tools. As the church rightfully proclaims how our faith connects to all of life, including our work, I am grateful for this volume by these distinguished scholars and practitioners addressing the unique challenges that technologists face every day, given the importance of their work that touches every aspect of our society. This is an extremely valuable resource connecting Christian theology and ethics to the pressing challenges of technology today."

Jason Thacker, author and chair of research in technology ethics at the Ethics and Religious Liberty Commission of the Southern Baptist Convention

"Brue, Schuurman, and VanderLeest have provided an excellent resource for contemporary students who may be designing technology, systems, or software, helping them to think about the implications of what they are designing. It covers broad biblical principles that technologists can bring to the workplace to help shape a design in a responsible manner. The authors also recognize that even the best and most well-intentioned designs can be subverted and become yet another idol in the lives of those who design and use them. This guidebook will help Christian engineers and computer scientists think more biblically about the technologies and systems they develop. My only wish is that it had been written when I was an undergraduate engineering student five decades ago!"

Dennis L. O'Neal, dean of the School of Engineering and Computer Science at Baylor University

"Reading the pages of this book brings me back to childhood dinner table discussions with my father, Dr. Charles Adams. Dad was a visionary in the area of technology and faith. He believed that what we did as engineers on this earth deeply mattered to God. I would recommend this book to anyone in a technical pursuit who is seeking to serve the Lord in their work. In my own experience, many Christians today feel they have to leave their faith outside the office. This is so tragic, as so much of life's meaning and blessing come from seeing the Lord's hand in our daily work. If you are a Christian engineer struggling to find meaning in your work, this book is for you."

Mike Adams, CEO of Adams Thermal Systems

A CHRISTIAN FIELD GUIDE TO TECHNOLOGY FOR ENGINEERS AND DESIGNERS

ETHAN J. BRUE

DEREK C. SCHUURMAN

STEVEN H. VANDERLEEST

Academic

An imprint of InterVarsity Press

Downers Grove, Illinois

InterVarsity Press
P.O. Box 1400, Downers Grove, IL 60515-1426
ivpress.com
email@ivpress.com

InterVarsity Press® is the book-publishing division of InterVarsity Christian Fellowship/USA®, a movement of students and faculty active on campus at hundreds of universities, colleges, and schools of nursing in the United States of America, and a member movement of the International Fellowship of Evangelical Students. For information about local and regional activities, visit intervarsity.org.

The publisher cannot verify the accuracy or functionality of website URLs used in this book beyond the date of publication.

Cover design and image composite: David Fassett
Interior design: Daniel van Loon
Images: stock market price display: © baona / iStock / Getty Images
> *blueprint grid lines: © belterz / E+ / Getty Images*
> *circular blue print drawing: © belterz / E+ / Getty Images*
> *clockwork cogwheels: © benedek / E+ / Getty Images*
> *computer component motherboard: © Creative Crop / Digital Vision / Getty Images*
> *periodic table: © demarco-media / iStock / Getty Images*
> *computer circuits chips: © MirageC / Moment / Getty Images*

ISBN 978-1-5140-0100-4 (print)
ISBN 978-1-5140-0101-1 (digital)

Printed in the United States of America ∞

InterVarsity Press is committed to ecological stewardship and to the conservation of natural resources in all our operations. This book was printed using sustainably sourced paper.

Library of Congress Cataloging-in-Publication Data
A catalog record for this book is available from the Library of Congress.

P	25	24	23	22	21	20	19	18	17	16	15	14	13	12	11	10	9	8	7	6	5	4	3	2	1
Y	37	36	35	34	33	32	31	30	29	28	27	26	25	24	23	22									

CONTENTS

PREFACE

Whether you are a student or a seasoned professional, many Christians working in technical areas struggle with engaging their faith in the world of technology. Perhaps you never thought about how your faith might inform your work in technology. Perhaps you feel a disconnect between your daily professional life and your Christian spiritual walk. The development of technology through science and engineering has always been a cultural activity with religious implications, but its direction is set by the human heart. Developing and using technology is one way we love God and our neighbor, and more fully witness to the gospel of Jesus Christ for the entire world.

The vision for this book is to provide a guide for Christian engineers and others working with technology to responsibly navigate today's technological terrain. A field guide is a resource that helps the reader identify things (usually plants or animals) in their natural environment. We hope this book serves as a field guide for readers interested in engineering and technology to identify and discern technology and its cultural environment. Furthermore, since all readers will be users of technology, and many will be practitioners, this book provides some principles and advice that we hope will be helpful to Christians who wish to be faithful and honor God in the technological aspects of their lives.

A field guide is not intended to be complete or comprehensive, since it is designed to accompany, not exhaustively inform, the traveler. Outfitted with this guide, explorers share the journey guided by the thoughts of those who have traveled a similar path. This field guide shares multiple voices, with varying perspectives about the terrain ahead, but always on the same journey of discipleship.

A field guide also provides a balanced perspective by noting challenges like endangered species and habitat pollution, but also depicting the beauty and wonder of the natural world. Here too, we hope to acknowledge the challenges arising from technology, but also point to the wonderful possibilities that come with technology and its ability to contribute to flourishing. Many books in philosophy and the social sciences provide wise warnings about the threats associated with technology, but we also want to ask the practical question: How then shall we engineer? Like other cultural pursuits such as politics, education, or the arts, technology has consequences if not pursued thoughtfully and responsibly. We want to avoid being what Neil Postman called, "one-eyed prophets," who dwell only on the problems arising from technology, unable to imagine the wonderful possibilities.[1] Rather than wringing our hands in despair, we want to explore how we as engineers and designers can responsibly pursue our technical vocations.

The three authors were motivated to write this book because we have each had similar struggles that we suspect many readers have. Each of the authors have served as professors in Christian colleges and universities but have also spent time in industry. Ethan Brue is an engineering professor at Dordt University and has worked as a project engineer in the energy and agricultural technology industries. Derek Schuurman is a computer science professor at Calvin University, but his academic training and industry experience are in electrical engineering. Steve VanderLeest is currently a computer engineer working in industry but previously served as a professor of engineering at Calvin University.

An article published in the journal *Christian Higher Education* surveyed faculty in Christian higher education and found that professors specializing in computer science, math, and engineering were the least likely to integrate faith into their teaching.[2] Our experience has shown that this is not due to lack of piety on behalf of Christian engineering professors, but rather the challenges of faith integration within certain disciplines. As authors, we understand these challenges firsthand; each of us has worked as an engineer in industry as well as taught in Christian higher education. By sharing what we have learned, our hope is to encourage readers to ask wise questions about technology, aspire to more noble purposes with technology, and live lives more consistent with our faith in the areas of technology.

We acknowledge that the topic of faith and technology is complex, and as authors we did not always agree on precisely what to include or not. In fact, the purpose of this book is to invite you to join the ongoing conversation and participate—exploring, discerning, and at times wrestling. Thankfully, none of us need to do this on our own. We can benefit from the insights from a large cloud of witnesses. We recognize that this book stands on the shoulders of many others, including the groundwork laid out in the 1986 book *Responsible Technology.*

You will notice multiple voices throughout this book. Chapter one was a collaborative effort. Ethan Brue developed chapters two and seven. Derek Schuurman crafted chapters five, six, eight, and ten, and Steve VanderLeest wrote chapters three, four, and nine. Throughout our writing of the chapters, conversations between all three authors provided different perspectives to challenge, shape, and encourage. We hope the presence of multiple voices encourages the reader to join our conversation as well. We have added reflection questions at the end of the book and a companion website to continue the conversation with related links, articles, and resources, which the interested reader can find at ivpress.com/fieldguide.

Part of this book manuscript was written during the onset of the coronavirus pandemic, a time when we were reminded of the blessings of technology to facilitate ongoing communication during a time of necessary physical distancing. But it was also a time in which we were reminded that we are not in control; that despite all our ingenuity we remain dependent on our heavenly Father. We are grateful to God for his grace and faithfulness which is also evident in the gift of technology.

Soli Deo Gloria

ACKNOWLEDGMENTS

The seed for this book was first planted when one of the authors, Derek Schuurman, spent the 2015–2016 school year as a visiting professor at Dordt University alongside Ethan Brue. During that time, Derek team-taught the capstone course for engineers and computer scientists at Dordt. The capstone course made use of the book *Responsible Technology* (edited by Stephen Monsma and published by Eerdmans in 1986), a book that is now over three decades old. At one point during Derek's time at Dordt, the suggestion was made to explore writing an update to that book. This was not the first time such an idea was proposed. In 1999, Steve VanderLeest had begun discussing such a book with Gayle Ermer of Calvin University and another Dordt engineering professor, the late Charles Adams, envisioning an approach that was more practical and a bit less philosophical than the earlier textbook. At that time, Charlie even mentioned a promising new faculty member at Dordt, Ethan Brue, who might contribute to the project. Due to other demands, the book project remained just an idea. Having originally met at conferences of the Christian Engineering Society, Derek and Ethan invited Steve to discuss and explore the idea for such a book. After a meeting at Dordt in March 2016, the three authors agreed to forge ahead, and the result is the book you hold in your hand.

The authors are grateful to many people for assistance and encouragement with this book. We are grateful to the Andreas Center for Christian Scholarship at Dordt University and the Calvin Center for Christian Scholarship at Calvin University for grants and funding in support of this project. Moreover, Derek Schuurman is grateful for the time made possible by the William Spoelhof Teacher-Scholar-in-Residence chair at Calvin University. The authors are also grateful to many individuals, particularly Sally Jongsma for her assistance in editing an early draft of this book. We are also grateful to those who provided feedback on an early manuscript, including Adam Blankespoor, Nick Breems, Mike Capozzoli, Rebecca Konyndyk DeYoung, Susan Felch, Monica Groenenboom, Fred Haan, Nathan Jen, Calvin Jongsma, Sally Jongsma, Arie Leegwater, Quentin Schultze, Kevin Timmer, Keith Vander Linden, Kevin Vander Meulen, Justin Vander Werff, Jeff Van Dorp, Nolan Van Gaalen, and Justin Voogt.

We are grateful to all the folks at IVP Academic, including two anonymous reviewers, and particularly our editor, Jon Boyd, for his sage feedback and his shared vision and enthusiasm for this book.

The authors are also grateful for mentors in our own lives—people who have encouraged us to develop as Christian engineers. The Dr. Van Wijs character in the last chapter is a composite of the many people who have been an encouragement to us. Our hope is that this book will be an encouragement to you and that it will stimulate further writing and conversation.

Finally, we would like to express our gratitude to each of our spouses, Donna Brue, Carina Schuurman, and Pam VanderLeest, for their love, encouragement, and support.

CHAPTER ONE

DREAMS TAKE FLIGHT

Not only cathedrals, but every great engineering work is an expression
of motivation and of purpose which cannot be divorced from religious
implications. . . . The age of cathedral building is long past. . . . But every
manmade structure, no matter how mundane, has a little bit of cathedral in it.[1]

Samuel Florman

Before heading off into the technological landscape, we begin our field guide with an examination of the underlying yearning to design technology. What drives our dreams and longing to invent? This chapter was written collaboratively by all three authors, reflecting insights from their own journeys to become engineers. The purpose of this chapter is to connect our hopes and dreams exhibited in technology to our ultimate hope in our sovereign God who is making all things new in Christ.

Two dreams. Same goal. Same technology. Different endings. Samuel Langley dreamed of a flying machine. So did Orville and Wilbur Wright. In fact, so did many others, stretching back to the artist-inventor Leonardo da Vinci in the thirteenth century. Of all the dreamers, inventors, and innovators through the ages, the story that ultimately gets tied to the invention of human flight tells us as much about our own hopes and dreams as it does about the transcendent imagination fueling the technology itself.

In the mythology of flight, the Wright brothers most often play the protagonists. However, in the late 1800s, if you were betting on the future of heavier-than-air powered flight, Samuel Langley would have had the best odds for being the inventor of the early aeroplane. He had the right education as a scientist, political capital, connections, and finances to make it happen. As a distinguished astronomer and head of the prestigious Smithsonian Institution, he had at his disposal seventy thousand dollars (nearly two million in today's dollars), of which the majority was taxpayer funded, to design, build, and test the first mechanically powered airplane. By contrast, with less than a thousand dollars of their own money, the Wright brothers were self-educated, self-funded, and self-motivated owners of a bicycle shop. This backdrop brings to mind our favorite storylines in which an unlikely underdog overcomes insurmountable odds to achieve success.

In the fall of 1903, Langley's scientifically engineered contraption, packing an impressive fifty-plus horsepower, made its long-awaited, well-publicized flight multiple times across the Potomac in Washington, DC, complete with a sizable entourage of reporters, scientists, and interested citizens. While the stage was impressive and the flight commendable, it was the dramatic landings that stole the show. The final and most spectacular landing occurred in December of 1903, morphing aeroplane into submarine and memorably landing a brave but fully chilled and drenched pilot, sputtering profanities, back on the riverbank. What crashed in the Potomac was more than just the flying mechanism. Equally damaged that day was the public's faith in institutional science, political power, and wealth. Meanwhile, a little over a week later, two bicycle mechanics who spent a mere four years of vacation time "playing with" technology were able to coax a mere twelve horsepower engine to lead them into history, witnessed by a handful of curious locals with surprisingly little drama.

Maybe we find the story of the airplane so intriguing because it embodies the rags-to-riches mythology that we want to be true. However, a closer reading of the Wright story reveals something deeper than the retelling of the American dream. The story reinforces the notion that in culture making, the visionary artist often eclipses the scientist. This may explain why today we remember Kitty Hawk as the site of the first flight and why the Wright brothers achieved a chapter in history while Langley only secured a footnote.

What drove the brothers Wright to dream about flying? Momentous technological change often grows from a deep yearning or belief. The Wright brothers' imagination was sparked at an early age after Milton Wright gave a toy flying device to his young sons, Orville and Wilbur, in 1878. It was perhaps an odd gift coming from a man of the cloth, as Milton was a bishop in the Church of the United Brethren in Christ. Little did their father know that his gift might inspire the boys to dream of creating the first powered aircraft that could be reliably controlled by a human pilot. Not all beliefs are articulated in doctrinal statements, some take shape in wood and metal.

The boys grew up in Dayton, Ohio, pursuing diverse interests in sports, music, nature, and mechanical devices. As teenagers, their curiosity drove them to build their own printing press and to later start a bicycle shop. Reading news stories about early attempts to fly reinvigorated their childhood dreams, and they began researching and experimenting to create their own airplane. Possibly taking their cues from their bicycle world, they reconceptualized the "problem of flight" not as getting into the air, but as giving us control when we got there. Like the experience of learning to ride a bike, our earliest failures occur when trying to turn, slow down, or stop. Driven by this quest to solve the flight control problem, they built models and prototype gliders to test and refine their design. They meticulously examined each failure and improved the design before ever attempting powered flight—all the while refining their skill at riding the airplane. They identified the Outer Banks of North Carolina as an ideal location for their final field experiments because of its frequently windy conditions. They set up camp in Kitty Hawk, being drawn back to this wilderness landscape not simply for its aerodynamic advantages, but also for the contest of wit and skill played out against an unpredictable foe of wind and weather.

Figure 1.1. Photo of the first powered, controlled, sustained airplane flight in history at Kitty Hawk, North Carolina, December 1903

Longing represented by dreams and imagination is an important factor in technology design. Today, most new technology originates in big companies, so we might mistakenly connect innovation to the economic motivations of corporate businesses and miss the role of play. One of the most striking features of the Wright brothers' story, though, is that their motivation arose not from some practical need but from a delight in tinkering and exploring. Their story demonstrates the power of play, imagination, and human creativity—driving them from the inauspicious printing press and bicycle shop in Dayton, Ohio, to flight tests in the boondocks of Kitty Hawk. Kitty Hawk was more akin to a rustic vacation camp than a science project, and the memoirs of their experiences on the Atlantic coast read more like poetry than a lab report. Orville and Wilbur were gymnasts, football fans, pond hockey players, bike riders, skate sharpeners, book lovers, naturalists, art connoisseurs, and musicians, as well as inventors. These diverse interests shaped their imaginations, and their curiosity drove them to design the world's first successful heavier-than-air engine-powered and pilot-controlled aircraft.

More important than the creation of the Wright Flyer itself, may be the creation of the Wright history. Langley and his institution of scientific predecessors may have, through their myriad of failures, done more for the ultimate future of aircraft design than the Wright brothers. Nevertheless, it is the Wright brothers that we more often choose to remember. Stepping back from the story, the reason may be bound up in our dreams. We also dream of the activity of creating: the adventure, the joy, the delight of exploration, free from the demands of our modern industrial machine. While only a few in the world are granted access to the well-educated and well-funded world of Langley, most of us can identify with the world of the Wright brothers. Their dreams of creating were as strong as their dreams of flying.

What drives humans to create technology? Some have suggested that invention is the result of Darwinian selection and that creating tools to survive is simply an evolutionary skill that developed over time, corroborating the old adage that necessity is the mother of invention. While some creative abilities that our ancestors employed to survive still endure, such as harnessing fire for energy, other creative endeavors are now directed toward crafting the arguably less essential jet skis and big screen televisions. Others have suggested that invention is a type of technological determinism, assuming that technological progress is inevitable, implying that engineers are compelled by some impersonal force. Still others have suggested that invention is driven primarily by consumerism and materialism, which creates the demand for what invention supplies.

Is invention only driven by the instinct for survival or the instinct of greed? Neither survival nor greed led us to the first flight at Kitty Hawk or the first majestic cathedral. They were driven by delight in creating something new and beautiful and noble. Technology always serves a purpose or seeks to achieve a goal by solving problems. The best technology delights us with intuitive melding of form and function, and in some ways this aesthetic makes it a product of not only science but also of art. Although calculation and logic are fundamental to modern technology, the development of technology is a creative activity. We do not calculate a new technology, we *design* it. In the end, our best technology is derived not from our base instincts but from our noblest dreams.

Yearning for something better stokes our imaginations to explore new possibilities and to envision a different reality. Imagination is our conscious dreaming. Our ability to create technology allows us to ponder new and better ways to achieve our goals, and even to conceive of new goals that build on new tools. We can better our lives, improve our community, advance our society, and care for our world with the devices we dream up and build.

Science fiction has long inspired the imaginations of modern technology developers. Written in 1865, Jules Verne's classic tale *From the Earth to the Moon* imagined a voyage to the moon a century before the first moon landing. Another classic Verne story, *Twenty Thousand Leagues Under the Sea*, describes an electrically powered submarine a decade before the first one was constructed. Prolific science fiction author Arthur C. Clarke described artificial satellites for communication more than a decade before the USSR launched the Sputnik satellite in 1957. Today satellite technology is so common that we usually drop the qualifying adjective "artificial." Martin Cooper, an AT&T engineer widely recognized as the father of the cell phone, mused that *Dick Tracy* comics depicting futuristic wrist-watch radios may have inspired his vision for mobile phones. The *Star Trek* television series may have also inspired various technologies, ranging from spaceship propulsion to medical imaging. In one of the halls of the Smithsonian National Air and Space Museum one can view a model of the starship *Enterprise* from *Star Trek* near a model of the lunar landing module, placing the dream of space travel next to the reality of it. Indeed, imagination often precedes invention.

DREAMING OF SPACE

One of my indelible childhood memories is the sight of a television image of the launch of Apollo 17 and the subsequent fuzzy black and white video of men walking on the moon.[2] It was late 1972 when I watched footage of Walter Cronkite's report on the Apollo program's final mission to the moon. My father let me stay up late one evening to watch the spectacular images of people visiting another world. I didn't know it at the time, but these missions to the moon were the culmination of centuries of human dreaming about flight, outer space, and celestial bodies. What I did know was that it sparked my own imagination. It was one of the experiences that led me to dream about my own vocation: first imagining myself as an astronaut, but eventually focusing more on technology and deciding to become an engineer.

Figure 1.2. Apollo 17 launch

The entire Apollo program was the culmination of a dream. In 1961, US President John F. Kennedy laid out an ambitious, mind-boggling goal:

> I believe that this nation should commit itself to achieving the goal, before this decade is out, of landing a man on the moon and returning him safely to the Earth. No single space project in this period will be more impressive to mankind, or more important

for the long-range exploration of space; and none will be so difficult or expensive to accomplish.

Tragically, Kennedy didn't live to see his challenge met. On July 21, 1969, astronaut Neil Armstrong stepped onto the surface of the moon. As he descended from the lunar landing module onto the surface of the moon, Armstrong said, "That's one small step for a man, one giant leap for mankind." People watching this historic moment could only imagine what leaps in space travel might lay ahead.

Like the explorers centuries ago that set out on uncharted waters to explore the unknown reaches of earth, many of those drawn to the original US space program were spurred by a dream of exploring the unknown reaches of space. Robert H. Goddard was the American scientist who built the first liquid-fueled rocket. He became fascinated by spaceflight after reading H. G. Wells's novel *The War of the Worlds* as a young man. The concept of interplanetary travel gripped his imagination, influencing his decision to study physics with a focus on developing rockets.

Wernher von Braun also dreamed of rockets and space travel. As a youth, he read a science fiction novel about two planets and grew up to become an aerospace engineer. However, his commitment to his dream of rockets eclipsed other concerns. His obsession led him to work on the V-2 rocket for Nazi Germany, even though his real interest was in space travel and not weapons of destruction. As the war came to a close, he arranged to surrender to American forces, hoping it would offer the best path for further rocket research. It did. He joined the US space program and is widely regarded as the inventor of the Saturn V rocket.

After retiring, Katherine Johnson would visit with aspiring students and recount her career at NASA, telling them to never give up on their dreams.[3] Johnson had joined NASA as a mathematician, calculating trajectories for spacecraft by hand, since this was in the 1950s, before the era of the electronic computer. She was the first woman to publish a technical report at the space agency.[4] Her story of success as a Black woman in an organization dominated by White men became the subject of a 2016 feature film, *Hidden Figures*.

Figure 1.3. Katherine Johnson at her desk at the NASA Langley Research Center, 1966

DREAMING OF COMMUNICATION

Dreams can also be spurred by tragedy. Samuel Morse, inventor of the telegraph, originally worked as a portrait artist. While working on a commissioned piece that took him far from home, his wife became seriously ill and died. He learned of her sickness and death too late to return home before she had already been buried. Heartbroken, he began to dream about a way to communicate that would be faster than sending a message via horseback. A conversation with a scientist about electromagnetism a few years later gave him an idea for building a message-sending machine. Over the next decade he worked on a device that included a transmitter, a receiver, a relay to amplify the voltage over long distances, and a code to translate the messages. He patented his ideas in 1840, claiming an invention for "Improvement in the Mode of Communicating Information by Signals by the Application of Electro-Magnetism." Some of the intricate details of his invention are depicted in figure 1.4 as they appeared in his patent application for the device. In May of 1844, Morse sent his first message by telegraph: "What hath God wrought?" He chose the content of his first communication from the Bible (Numbers 23:23 KJV).

Centuries before Morse, Johannes Gutenberg invented the movable metal type printing press, choosing the Bible as his first publication.

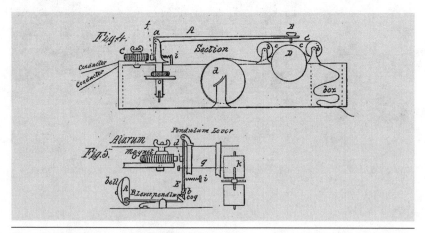

Figure 1.4. Two diagrams from the Morse Telegraph patent application

The Gutenberg Bible, as it came to be known, was the first mass-produced print book. Over 150 copies were printed around 1440, and the remaining forty-nine copies that have survived are considered some of the most valuable books in the world today. Later inventors were not as clever or biblical with their inaugural messages. A couple of decades after Morse, Alexander Graham Bell used the first telephone in 1876 to summon his assistant: "Mr. Watson, come here, I want to see you." A century later, Ray Tomlinson sent the first email message in 1971, seeking new ways to communicate with others over a distance.

What do these messages have in common? They marked the moment the inventor's dreams took flight and became reality. Each new communication technology allows people to connect with other people in a new way, forming new relationships at a distance not previously possible. Connecting might sound like a technical term, but it is a human term that describes the forming of a bridge between two people, linking them in relationship. It captures the ultimate and proper goal for technology—to help us be more fully human in relationship to each other and to God.

Imagination shapes technology, but technology can also shape our imaginations. At the turn of the previous century, theologian and statesman Abraham Kuyper reflected on radio communications in his devotional *Near Unto God*. He suggested that technology can "help us understand our God in ways that our imaginations couldn't as easily register before."[5] He observed that radio communication "helps us visualize how it is that the Lord, whose throne we say is in heaven, can pick up even a whispered prayer in the silence of our bedrooms." He continued, "The radio is thus even more of a gift of God—not only does it expand our world, it also helps us understand more about God's power."[6]

DREAMING OF LIGHT

Thomas Edison's creativity led to hundreds of inventions and thousands of patents. His dreams sometimes required weeks or even years of experimentation and hard work before they became reality. His dream of creating a better light bulb by improving on the earliest versions of electric lighting was not realized easily—he made over one thousand attempts before discovering the right filament material. It took more than a year of further effort before he produced the first commercially viable incandescent bulb. When he finally came up with a working, reliable device, he patented the idea.

His dream was not just a better light bulb, but rather the transformation of the entire urban way of life with a complete direct current generation and distribution system. Edison envisioned the cultural transformation of homes and industries with electric power. His innovation extended well beyond a single technology. His creations encompassed all components of the system, from generators, to meters, to conductors. A new electrified community was perhaps Edison's grandest dream and greatest invention.

Edison, Morse, the Wright brothers, and many other inventors have turned dreams into reality. Although they may all have had an extraordinary creative genius, all humans are endowed with creativity.

USERS DREAM TOO

Inventors and science fiction writers are often the first to imagine a new technology, but users, inspired and enabled by the new inventions, begin to dream too. New technologies often make a big splash in society, causing excitement, wonder, fear, and amazement. Consider the scene on the street circa 1900 as people saw the first horseless carriage drive past, experiencing astonishment, delight, and perhaps fear. As a society, we exhibit a similar range of reactions to new inventions, such as our reaction to autonomous automobiles that are not only horseless but driverless. Whether our response is positive, negative, or mixed, it seems we inevitably incorporate new technologies into our daily lives. Right now, many of us are probably exploring new devices or tech tools, perhaps an app, that friends or relatives have not yet tried. Some may be skeptical, doubting its usefulness or concerned that it will change their lives for the worse. Others might be favorably disposed toward the new device but cannot afford it. When new technology shows up, we are all forced to evaluate it and choose whether to adopt it. How can we make wise and faithful choices about the technology we adopt?

Imaginative users not only adopt technologies, but they often have further thoughts about how to adapt or modify them for uses beyond what the inventor envisioned. For example, although monks may have invented the clock to artistically depict the God-ordained movement of the heavenly spheres, this mechanism was soon used to gather them for prayer with more precise regularity. As the precision of mechanical time extended beyond what an hourglass or sundial could provide, it ushered in an era where tools to measure time became common. Individual users imagined uses far beyond marking prayer times. Modern watch-wearers use clocks to monitor laborers, set mealtimes, synchronize meetings, and more. Gradually, time zones were established, and with the ability to more precisely measure time, deadlines could be specified to the minute. The clock changed society, illustrating the notion that we shape our technology, but our technology also shapes us.

FAITH UNDERLIES ALL TECHNOLOGY DREAMS

Technology surrounds us more than we realize. While we easily notice the latest device as high-tech, we often fail to recognize the pervasive presence of older technologies because they have faded into the background of our conscious observation. Computer scientist Alan Kay noticed this fading: "Technology is anything that wasn't around when you were born." For example, the worship director at one of our churches often mentions that she hopes the technology will run smoothly during the worship service. By "technology" she means the computer and projector displaying the song words on the screen, the audio system that allows the musicians to hear each other with good balance, the video shown part way through the service, and the wireless microphone for the preacher. She probably doesn't think about all the other technologies that make up the service: the devices invented in previous ages that aid our formal worship like the piano, central heating, grape juice for Communion, architectural technology, carpeting, electric lights, printed Bibles, paved parking lots, and concrete steps at the entrance. All these technologies have faded into the background so that we hardly notice them unless they fail in some way. Similarly, you might recognize your smartphone as high-tech but fail to notice older technologies you are wearing right now, like your eyeglasses, stain-resistant pants, and even the zippers on your clothing. Noticeable or not, technology often changes the way we live, interact, work, play, and even worship.

The impulse to create technology is fueled by our dreams and our imagination. "In dreams begins responsibility," wrote the poet William Butler Yeats. He may not have been thinking of technology, but he understood that yearnings hint at responsibility and ultimate purposes. Technology has deeply religious roots. Our yearnings and imagination reflect personal values shaped by desires and longings within our hearts. They are glimpses of the ultimate dream: the longing to see the new creation and the new Jerusalem. Technology grows out of our human character endowed with the image of the Creator. It grows out of biblical callings to care for the creation, to love God, and to love our neighbor. Done well, our technological dreams become reality in building God's kingdom. Done poorly, our technological dreams become nightmares that pervert creation and harm our neighbor.

Most dreams fade away, but those shaped by the Spirit of God do not. This creative spirit is poured out on engineers who continue to see visions and dream dreams, prophetically pointing to a coming kingdom. The Christian faith does not restrain engineers from having extravagant ideas but encourages us to imagine what God desires. As Christians, our dreams must be animated by the biblical story, and our hearts must be tuned to a vision of God's coming kingdom. Christian engineers need to equip themselves with more than technical competency; they need to cultivate an imagination for how things ought to be. That imagination must be shaped by faith in Jesus Christ, in whom and through whom and for whom all things exist.

There is more to the story of technology than simply dreaming. Our design and use of technology start with dreaming but are also influenced by a variety of other factors. Achieving dreams consistent with our Christian faith in the complex context of real life is the subject of the rest of this field guide. We begin in chapter two with the Scriptures by reviewing the biblical story and describing how it can inform our understanding of technology. Chapter three dispels the myth that technology is objectively neutral in order to make the case that we must exercise responsibility. As an aid to wise design, chapter four introduces one of the key concepts of the book: design norms for responsibly creating and evaluating technology. While the norms from this chapter provide a framework for guiding God-honoring designs, chapter five looks at how norms and virtues can inform and expand the topic of professional ethics. Chapter six examines the impact of sin on our technological work and products. Chapter seven provides a historical perspective, encouraging us to continually "zoom out" to see and imagine the big picture when immersed in the minutiae of a complex project. Leaping forward from the lessons of history, in chapter eight we hope that by pointing our design efforts to our broader eschatological hope, we can avoid undue utopianism or pessimism. In chapter nine we demonstrate technology design as a legitimate Christian calling, making the case that an engineer or scientist need not quit her job in order to serve Christ fully and faithfully. We end the book by addressing some concerns you might have about how one practically lives out their faith in the design of technology. Chapter ten takes the form of letters to a young engineer, giving practical real-world examples of how the ideas of this book become real—where the rubber meets the road.

CHAPTER TWO

A SURVEY OF TECHNOLOGY
AND THE BIBLICAL STORY

Story is the primary way in which the revelation of God
is given to us. The Holy Spirit's literary genre of choice is
story. . . . To get this revelation right, we enter the story.[1]
EUGENE PETERSON

Our first field study of technology will begin at a high elevation to see how the full sweep of the biblical narrative informs our understanding of technology. Ethan Brue will be our guide as we survey the entire biblical landscape with an eye for technology across this expanse.

On Christmas Day, December 25, 1968, at 9:30 p.m. EST, an alignment of three world-changing technological innovations made an isolated astronomical event viewable with the naked eye to humans in virtually every major city on earth. A spacecraft, television, and printing presses allowed people everywhere to be entertained by a crew of technical experts (a.k.a. astronauts) in the tight confines of the Apollo 8 spacecraft reading responsively from the first chapter of Genesis.

Astronaut William Anders: In the beginning God created the heaven and the earth. And the earth was without form, and void; and darkness was upon the face of the deep. And the Spirit of God moved upon the face of the waters. And God said, "Let there be light": and there was light. And God saw the light, that it was good: and God divided the light from the darkness.

Astronaut James Lovell: And God called the light Day, and the darkness he called Night. And the evening and the morning were the first day. And God said, "Let there be a firmament in the midst of the waters, and let it divide the waters from the waters." And God made the firmament, and divided the waters which were under the firmament from the waters which were above the firmament: and it was so. And God called the firmament Heaven. And the evening and the morning were the second day.

Astronaut Frank Borman: And God said, "Let the waters under the heaven be gathered together unto one place, and let the dry land appear": and it was so. And God called the dry land Earth; and the gathering together of the waters called the Seas: and God saw that it was good.

Upon seeing a moon's-eye view of our planet from an angle never before seen, it seems natural that humans would break forth in a hymn of praise and awe to its Creator. People often have similar responses to seeing a sculpture, a painting, a skyscraper, a bridge, or a home. We are seldom satisfied with just seeing the work, we want to know the artisan: Who made this? We have an irresistible desire to reach deeper into the creation and know more of the Creator. However, there is another layer to this story from 1968. The astronauts were not merely celebrating the first amazing image of the earth from space—we had already received a variety of snapshots for over twenty years. They were sharing with us the experience of an extraterrestrial viewing of our world, mediated by the marvel of a new technological communication medium (television), and expressing this technologically recorded ode to the Creator from the belly of a technological exoskeleton (the Apollo orbiter). This story celebrates both creation and its created creators. The God who designed the expansive cosmos that surrounds us also engineered the creative capacities within us. While God called out the entire cosmos from nothing, he enables engineers to call out the spacecraft, television, and printed version of Scripture from what he has made.

SHALOM is the state of active harmony that is realized only when humanity acknowledges its true rest and delight in our submission to God, service to humanity, and care for the creation.

The astronauts only read the creation account of the first two and a half days, an abbreviated reading from the most popular abridged version that describes a six-day creation narrative. However, the complete creation story involves seven days, not six. In the first six days God created everything in integral dependency, with every created thing dependent on all other creatures and looking to God for sustenance. Then, on the seventh day, he gives the creation its direction. He actively rests with his creation and sets into motion a celebration of creative potential with purpose. The seventh day is the never-ending invitation for creation to live and work in right relationship with God, fellow humans, and the rest of creation. The biblical term for this condition of harmony is *shalom*.[2] Service and care are what bind creation together and enable it to offer unified praise.

The astronauts' journey into space was possible because of the other six-day creations: the hydrogen and oxygen obediently hurling them into orbit; the mosaic of metallic crystals, plant fibers, mineral castings, and polymer composites graciously separating the humans from the suffocating darkness of space; not to mention the staff at mission control, the network commentators, the television viewers, the tax paying public, the schools, the families, the government all fulfilling their roles of service—the list is unending. Whether the fragile astronauts gently buoyed by finely tuned gravitational currents acknowledged it or not, the entire enterprise was dependent on God. They were celebrating the act of creating, which is eternally bound up in the creation story. The story of the space race, as with all human activities, is mixed with prideful ambitions, misplaced faith, and political maneuvering at individual, institutional, and national levels. Still, we should not miss the goodness of the technological unfolding that comes when humans seek to "fill the earth" and bring out its potential. As humans create, fill, and enjoy creation, they imitate the Creator God who, with us, celebrates all obedient unfolding of his creation.

The first chapter illustrates that technology gets its direction from our deepest longings and dreams. This chapter will summarize how Scripture informs our Christian hope and imagination in relation to technology. The biblical story reveals God's deepest longing to make all things right, restore his creation, and dwell with his people.

Despite the reputation of engineering as a rigorous and demanding field, its challenge is not in its immense complexity but in its mind-stretching simplicity. Engineering students often spend so much time trying to master all of the cursory concepts that the central ordering principle escapes them. In many engineering and science classes, the entire curriculum is built on learning the same three or four lawful constructs, albeit viewed from different angles or described by a different lexicon. For example, the entire course of fluid dynamics can be distilled into applications and derivatives of three core physical concepts: mass conservation, energy conservation, and momentum principles. Some texts even reduce this into a single equation for all properties and derive all three principles using the same mathematical relation. Additional content does not replace what we already know; it gives us a *fuller* picture.

> **SABBATH** is God's final act in the creation narrative. God dwells with his creation to bring it to its intended fullness in the unending seventh day. It is the culmination of God's redemptive plan, through Jesus Christ, to bring all things into eternal rest.

The same thing is true for the canon of Scripture. The rudimentary themes of Scripture are all found in Genesis 1–3: God the Creator of everything good will dwell (sabbath) with his creation, and he will keep his covenant of love with us by entering our world as Immanuel, "God with us."[3] The good news is that simple. But thanks be to God, he does not leave us with this perplexing simplicity. He reveals to us again and again what it looks like. Arguably, one of the most concise and yet comprehensive creation accounts is sung by the apostle Paul in his letter to the Colossians:

> The Son [Christ] is the image of the invisible God, the firstborn over all creation. For in him *all things* were created: things in heaven and on earth, visible and invisible, whether thrones or powers or rulers or authorities; *all things* have been created through him and for him. He is before *all things*, and in him *all things* hold together. And he is the head of the body, the church; he is the beginning and the firstborn from among the dead, so that in everything he might have the supremacy. For God was pleased to have *all his fullness dwell in him, and through him to reconcile to himself all things*, whether things on earth or things in heaven, by making peace through his blood, shed on the cross. (Colossians 1:15-20, emphasis added)

This is the biblical story summarized in its entirety. God creates *all things*—time, matter, energy, sensations, inferential relations, turnips, maple leaves, limestone, and potassium chloride. But there's more. He not only created it all, he is willing to offer his own lifeblood to bring his creation back from death into its intended fullness. God will continue his creative and re-creative work until all things are brought back into proper relation and all things enter his eternal sabbath. Even after humanity snubs God's invitation to sabbath with him, God delivers his chosen people from slavery and provides weekly, monthly, and yearly feasts and observances that follow seventh-day rhythms (e.g., land sabbaths and the Year of Jubilee). These foreshadow the fulfillment of the eternal sabbath in the person of Jesus Christ, who summarizes his entire mission as the Jubilee-in-the-flesh, quoting from Isaiah 61:1-2 and declaring,

"The Spirit of the Lord is on me, because he has anointed me to proclaim good news to the poor. He has sent me to proclaim freedom for the prisoners and recovery of sight for the blind, to set the oppressed free, to proclaim the year of the Lord's favor." Then he rolled up the scroll, gave it back to the attendant and sat down. The eyes of everyone in the synagogue were fastened on him. He began by saying to them, "Today this scripture is fulfilled in your hearing." (Luke 4:18-21)

The Bible's many chapters are bookended by the first two chapters of Genesis—telling of the Creator God putting all things into proper relationship and preparing for endless celebration with his creation—and the last two chapters of the book of Revelation—imaginatively retelling of the re-creating God putting all things into proper relationship and preparing for endless celebration with his restored creation. All chapters in between detail the bridges God builds at an unfathomable cost to make this happen. Jesus, Immanuel (God with us), is the final fulfillment of the ultimate Sabbath. In this foundational sense, there is nothing new in Scripture from beginning to end; the story does not change or get superseded, but it does get continually filled anew.

It is impossible to highlight all that Scripture illumines regarding technology and engineering in one chapter of this book. But the first chapters of Genesis frame or allude to a theme that fits the entire biblical narrative: creation-fall-redemption summarizes how God acts and how this covenant with his creation relates directly to our work as engineers and technologists.

THE CREATION AND THE PROCESS
OF EX NIHILO DESIGN

It is good. It is good. It is good. God is not afraid of being purposefully redundant as he makes a world, step by step.[4] Again and again and again he acts in repetition, until a level of fullness can be imagined. In God's divine design, the earth and all that is in it was good from the beginning, and it was "very good" once all the relationships were in place. The best designs are always iterative. As engineers, the stuff we responsibly study, name, form, and shape from the cosmos is also good. It is not just the matter and organisms of our world but also the intended relationships (human to human, human to creation, everything to God) that are very good. It follows that, when humans respond in obedience to God's call to name his creatures and bring fullness to the earth with our science and engineering, it is very good.

There are only two types of beings: God and creature. And since there is only one God (Deuteronomy 6:4, "Hear O Israel: The LORD our God, the LORD is one"), everything else must be "creature." A creature is any being that is created. This includes atoms, molecules, logic, numbers, space, mass, and time, in addition to plants, animals, humans, and everything else that shares our world. Nothing in this world exists that was not created. All things were created by God and are sustained by him. Nothing in creation is self-sufficient. Everything is dependent on the Creator. It is only God himself that is nondependent. God just *is*—or as he declares, "I am."

The creaturely experience of order is too often taken for granted. Read Genesis 1 again. What is essential and necessary for plants to exist? Contrary to conclusions drawn by many ancient and modern theorists, albeit for different reasons, it is not the sun—it is only God. As God adds humans to the mix and steps back to arrange all the pieces of his creating, he graciously requires *all* creatures (like the sun and daffodils and water and humans) to need each other. It was all good, but it wasn't "very good" until he finished it and put all creatures in appropriate service to one another. It is this lawful servanthood that enables us to describe biological relationships, such as photosynthesis. Contrary to our modern sensibilities, it is our *lack* of independence as creatures that in the end makes all things *very good*. Humans were intrinsically designed to live in inseparable communion with their Creator. Apart from God, we are always less than what we were created to be.

Many people assume that one of the goals of engineering is to make us more self-sufficient and less dependent. Advertisers and marketing executives want us to feel this kind of relationship to technology, even though a strong case can be made that we are more dependent today on more people for survival than we ever were in our preindustrial past. Ironically, this technological dependence on creation and each other is often hidden from us.[5] Being dependent on one another is not a problem for Christ-followers, who do not view self-sufficiency and independence as the end goal. We were created for non-self-sufficiency. A central purpose of engineering and technology, then, should be to enable all creatures to freely honor, serve, and depend on each other, as a means of becoming fully what we were created to be.

God remains exclusively God, and creatures remain exclusively creatures, but being transcendent (distinct from creation) does not mean God is not immanent (engaged with creation). He is both fully above creation and acting within it. Christ, as the image of the unseen God through the enlivening power of the Spirit, creates all things, holds all things, sustains all things, and reconciles all things to himself in order to dwell with them again. All things find unity in Christ—everything is oriented to this center (Colossians 1:17-20).

Creation, like a vector, has a twofold character: magnitude and direction. It was designed to contain a vast array of interdependent things with everything pointing to God. All things have worth and all things serve.

> Your word, LORD, is eternal;
>   it stands firm in the heavens.
> Your faithfulness continues through all generations;
>   you established the earth, and it endures.
> Your laws endure to this day,
>   for all things serve you. (Psalm 119:89-91)

All creatures find their purpose in extending honor and hospitality to God and to one another.[6] John Dyer creatively uses a programming metaphor to describe this twofold character of being (hardware) and purpose (software), recounting that

> To the stars he gives the job of separating day and night and marking out the seasons. To the plants he gives the job of sprouting fruit and seed. To the fish—what the Hebrew literally calls "the swimming things"—God simply says "swim." And to the birds—what the Hebrew literally calls "the flying things"—he says "fly." During those first six days, every plant and animal received a place and a function within God's world. What they are made to do—shining, spouting, swimming, and flying—in part defines what they are.[7]

Dyer helps us to reflect on how all of creation is defined, not only by what a creature is but by what it does. Creation serves God by doing what it was made to do. It was programmed to serve and praise, and that function cannot be muted. If you succeed in getting the disciples and children to stop shouting hosannas, you will only hear the rocks all the louder (Luke 19:37-40). The rocks have been praising the Lord of creation since at least the creation of time, if not before. The only way you can avoid "rock music" as an engineer who studies and works with creation is to have your ears and eyes closed. This is what John Calvin calls the "divine wisdom displayed for all who see."

> There are innumerable evidences both in heaven and on earth that declare his wonderful wisdom; not only those more recondite matters for closer observation of which astronomy, medicine, and all natural science are intended, but also those which thrust themselves upon the sight of even the untutored . . . so that they cannot open their eyes without being compelled to witness them . . . and [see] more than enough of God's workmanship in his creation to break forth in the admiration of the artificer.[8]

This is how the nonhuman choir of creatures works, day after day. However, human creatures can make noise rather than song. This is often the "sound" of sin. In music, a note in the wrong place can ruin what the original piece was written to be. In computer programming, a line of code that sends the routine in an errant direction is called a bug. We have many words to describe engineering that misses the mark, but we certainly could call it *antinormative,* a term that describes moving in a direction contrary to the biblical norms we will discuss further in chapter four.

THE CREATION AND THE ORIGINAL GREAT COMMISSION

For engineers, our first days on the job are usually spent familiarizing ourselves with organizational charts and job descriptions. Genesis 1 and other parts of Scripture also outline our foundational organizational responsibilities and job descriptions. In the voice of the psalmist,

> When I consider your heavens,
>   the work of your fingers,
> the moon and the stars,
>   which you have set in place,
> what is mankind that you are mindful of them,
>   human beings that you care for them?
> You have made them a little lower than the angels
>   and crowned them with glory and honor.
> You made them rulers over the works of your hands;
>   you put everything under their feet:
> all flocks and herds,
>   and the animals of the wild,
> the birds in the sky,
>   and the fish in the sea,
>   all that swim the paths of the seas.
> LORD, our Lord,
>   how majestic is your name in all the earth! (Psalm 8:3-9)

Here God describes for us the unique relationships he has established between his creatures. Humans are uniquely called to the freedom of obedient servanthood. This capacity to respond or refuse to respond is something only people are given. As human stewards, we are tasked with providing the space, energy, and time that all creatures need to aptly serve their co-created creatures, and in doing so, serve their Creator.

> So God created mankind in his own image, in the image of God he created them; male and female he created them. God blessed them and said to them, "Be fruitful and increase in number; fill the earth and subdue it. Rule over the fish in the sea and the birds in the sky and over every living creature that moves on the ground." (Genesis 1:27-28)

Being a steward gives humans the twofold task of responsibly caring for the creation and imitating the Creator. This is, in part, what it means to be created in the image of God. God breathes his creative spirit into dust and humans become living souls, and the cosmic artisan gives them their own workshop—a garden to unfold and enfold, enable and protect, cultivate and preserve. It can be helpful to think of this original Great Commission (sometimes called the "cultural mandate") as an internship. God never relinquishes his rule over all things, but he does invite humans to co-rule and co-keep. In one image of collaboration from the creation story, God brought creatures to man so that man could name them (Genesis 2:19). According to Nancy Pearcey and Charles Thaxton, "It was idiomatic in Hebrew that to name something is to assert mastery over it."[9] In the world of the ancient Israelites, the task of naming also implies a call to discover a facet of the creation and to determine how it functions. Technological development is one type of "naming" that calls us to both discover and steward creation by cultivating and harnessing creation's potential for the good of all things.

For astronauts to read from Genesis in a human-created orbiting metallic skin (AD 1968) fits the Augustinian tradition (AD 425) of reading from Scripture the creation story of our human-creating.

For, quite apart from the arts of living well and attaining eternal happiness—which are called virtues and are given only by the grace of God, which is in Christ, to the sons of the promise and of the kingdom—there are the many and wonderful arts discovered by human ingenuity, some serving our needs, some serving our pleasures. Even when it turns its desire to superfluous or, worse, to dangerous and harmful things, this extraordinary power of mind and reason shows what a great good it has by virtue of its nature, the good that enables it to discover, to learn, and to practice such arts. What wonderful, what astonishing heights human industry has reached in producing clothing and buildings! What progress it has made in agriculture and navigation! . . . What skill in measurement and number! How acutely it has come to grasp the motions and order of the stars! What enormous knowledge it has accumulated about the things of this world! . . . The creator of this wonderful nature is none other than the supreme God, who governs and guides all he created.[10]

Augustine does not feel the need to develop the argument that humans are called into technological pursuits. The tone of his praise is implication enough, and the allusion to the repeated refrain "and it was good" is apparent. In the beginning, God not only established our role as technologists, amidst all the other roles of culture making, he also modeled what this human-creating should look like.[11] God rested with his creation. This final day of creation has no beginning or end. As the only day without evening and morning limits, it defines the trajectory of all the other days. Genesis 2:15 states that God "put" or literally "rested" humans in the garden so that their "caring for" and "working with" would be one and the same. God intends that all creation finds its celebratory meaning in him. His desire is to dwell with his creatures, and along with his creatures to fill the whole earth with his glory.[12]

> The **CULTURAL MANDATE** is God's original com-
> mission to humanity in Genesis 1:26-28 and 2:15 to fill,
> rule over, care for, and keep the creation. It is our call
> to be culture makers that enable all of creation to
> flourish and reach its full God-glorifying potential.

Zechariah 14, set in the context of the Feast of Tabernacles or the Feast of Booths, illustrates how the Sabbath and Jubilee rhythm is an ongoing celebration of dependency and the joy-filled hope of all things oriented to God. The initial intent of the Sabbath (this final, unending day of creation) was that on the Lord's Day, "HOLY TO THE LORD will be inscribed on the bells of the horses, and the cooking pots in the LORD's house will be like the sacred bowls in front of the altar. Every pot in Jerusalem and Judah will be holy to the LORD Almighty" (Zechariah 14:20-21). This divine inscription will be stamped on everything, right down to the ordinary things we fabricate and engineer. The goal and purpose of creation was not efficient productivity but celebrative wholeness. Everything will find its fullness and unity in God's sabbath intent. Unfortunately, this goal is often absent from the predominant economic, social, political, and technological models we have adopted in overdeveloped nations, resulting in perpetual futility.

Augustine also points us to another biblical theme in the first two chapters of Genesis: abundant variety is God's trademark. The Creator God is and always will be a prodigal God.[13] He overdesigns everything. He completely abandons economies of scale and the efficacy of standardization. Interchangeability and uniformity are not part of his industrial vocabulary. He is unapologetically inefficient in using every potential color, shape, size, skill, and ability as he creates and recreates. God loves diversity, and he creates everything according to its kind. In fact, this radical differentiation, far from fracturing the world, becomes the resin that binds creation together. Every unique creature in every corner of the world has an irreducible task and irreplaceable role as ordained by God, who holds all things together. And at the apex of God's creative work and his creation is worship. Worship is the focal point of the creation narrative as told from Genesis to Revelation. Worship is not designed to be efficient, effective, or productive. All peoples, all nations, and all creatures worship because God deserves it (Psalm 148). Praise the Lord!

The technological rule of thumb we draw from this is that God's creation cannot be understood or described by a single aspect of our experience.[14] Our engineering ought to recognize that creation is held together as one and is at the same time irreducibly diverse.[15] For example, if we try as humans to create an object with only "energy" or only "economics" in mind and fail to notice either the "social" or "biotic" hues of creation, we will end up with a monochromatic or invasive technology that is as disadvantageous as a broken thumb.

THE FALL AND THE INTRODUCTION TO DESIGN FAILURE

The immense beauty that emerges from reading the first two chapters of Genesis soon turns repugnant in Genesis 3.[16] Something goes very wrong in the garden. Humans, in arguably one of the first biotechnological experiments, consume fruit which was deceptively marketed with the promise of giving them indestructible bodies and infallible minds. We failed to trust. We believed a lie. We threw away our lifeline and attempted to become something other than what we were created to be. The results have been catastrophic. Conflict emerges; rivalry is invented. Sibling is pitted against sibling, man against woman, nation against nation, humans against nature—it's the dog-eat-dog world we all know too well. Genesis 3 and 4 tell us that for those who live with their eyes and ears open in the world there is nothing new "under the sun" that we have not already experienced. The engineer-philosopher-teacher from Ecclesiastes says it this way:

I undertook great projects: I built houses for myself and planted vineyards. I made gardens and parks and planted all kinds of fruit trees in them. I made reservoirs to water groves of flourishing trees. . . . Yet when I surveyed all that my hands had done and what I had toiled to achieve, everything was meaningless, a chasing after the wind; nothing was gained under the sun. . . . So I hated life, because the work that is done under the sun was grievous to me. All of it is meaningless, a chasing after the wind. . . . A person can do nothing better than to eat and drink and find satisfaction in their own toil. This too, I see, is from the hand of God, for without him, who can eat or find enjoyment? (Ecclesiastes 2:4-25)

The teacher tells us something we already know. Ordinary, everyday life can seem so meaningless, but it can also seem good. Still, "toiling against" rather than "serving together," keeps life from being very good—at least for now.

Humans are creative. Sin and evil, in contrast, have always been creatively bankrupt. Sin can only take what is created good (technology, money, sex, work, power, etc.) and cobble together a crude imitation of what God created. It can only disfigure and twist the stuff of creation into some type of grotesque lookalike. In fact, that is the definition of "falling into sin" according to Scripture. Serving only the creature rather than the Creator, we engineer a system or gadget of silicon, chips, and codes, large enough to secure our desired market share. Then we heft our cumbersome idol onto the throne—and try to have an awkward one-way conversation with it. We begin to act like it. We even try to serve it.

Not to us, LORD, not to us
    but to your name be the glory,
    because of your love and faithfulness.
Why do the nations say,
    "Where is their God?"
Our God is in heaven;
    he does whatever pleases him.

But their idols are silver and gold,
> made by human hands.
They have mouths, but cannot speak,
> eyes, but cannot see.
They have ears, but cannot hear,
> noses, but cannot smell.
They have hands, but cannot feel,
> feet, but cannot walk,
> nor can they utter a sound with their throats.
Those who make them will be like them,
> and so will all who trust in them. (Psalm 115:1-8)

The words of the "nations" almost sound like the serpent in the beginning—offering a fraction of truth in the form of a lie. The truth is that we will not be "like God," but we will certainly become "like our gods," because all who make idols will become like them.

To better understand this point, let's look at artificial intelligence today. Too often Christian thinkers engage an issue without first dethroning the cultural idol—that computers will one day think like humans. If you are predisposed to believe that being human is rooted only in our analytical capacity or our reason, then it only takes a small step to envision and desire a future in which computers will one day be like humans. This dream or nightmare (depending on your perspective) presupposes that we agree about what it is to think. Instead of discerning theories of psychology guided by biblical revelation regarding what is central to humanity, we develop theories that assume that human cognition is a product of biophysical laws, stimulus-response events, or social conditions, all of which can be conveniently modeled, mapped, or predicted using algorithmic approaches.[17] What we ultimately arrive at is not a conclusion that computers can think like humans but that humans have capitulated to thinking algorithmically and statistically. Voluntarily or involuntarily, we have allowed what it means to be human to be confined to the boundaries of a new "Boolean reality."[18] In the end, computers don't think like humans; rather, humans more often succumb to thinking like computers. Those who make them will be like them.

We have become very efficient at idol making, and it is no longer dependent on the availability of silver and gold. We have recognized that by first declaring something "unconditionally non-dependently real," we can build our theorizing and serving on that creature.[19] In today's science-, technology-, engineering-, and math-driven culture, we make idols of data or matter or energy. This distorts not only our scientific theories, but also the engineering procedures that are inspired by such theories. For example, by starting with the assumption that the world is a great machine (the most common form of naturalism), we end up with theories that are mechanistic and designs that obey and appease this mechanistic deity. It doesn't matter whether you set out to make an idol, or you set out to worship an idol, or you simply adopt "common practices" of our world; with this mechanistic starting point, the final product will be an idol.

Let's take a moment to reiterate an important point: idol makers always take something that has a God-ordained role and purpose and ask it to do what only God can do. Isaiah 44 illustrates this in satirical prose as it describes an engineer who, with gifted planning and precision, renders and machines a 3D prototype from a composite biomaterial and then declares, "Save me, you are my god." It then juxtaposes a picture of the engineer using the same chunk of wood to build a crackling fire, complete with the savory smells of roasted meat, satisfying him with both warmth and sustenance. The point is that wood is created; therefore, it is good. Wood has thousands of ways it faithfully serves birds, insects, soil, water, and humanity with energy and food. Sinful humans blindly ignore the obvious. Instead of acknowledging the rightful place of creation's diverse elements, we foolishly try to give them authority they were never meant to have. It doesn't matter whether we are dealing with wood, energy, mathematics, silicon, fire, dance, programming, aggression, politics, manufacturing, or competition, the antidote to the poison of idolatry is to recognize that all things depend on God.[20]

Idolatry is always an impotent declaration of our human independence. Where our God (or our god) is, our trust will follow. Independent and alone in a broken world, we have limited options and so we easily turn to technological systems and artifacts for hope and comfort. There is an alternative way to live as human creatures, however. It requires asserting that "I am not my own, but belong, body and soul, in life and in death, to my faithful savior Jesus Christ," who alone can "set me free from the tyranny of the devil."[21]

Human creatures are at the religious center of creation; we are the conduit for God's salvation plan in our temporal world. It is through humanity that God, through the person of Jesus Christ, mediates his redemptive plan. When human hearts turn away from God, the rest of the creation suffers. Nothing remains untainted. There are no longer sanctuaries. No person, place, or thing in creation can be in the "all good" category anymore. Every individual, but also all families, churches, schools, organizations, will be infected with sin and its effects. Sin muddles it all.

REDEMPTION: REDESIGN AS RE-CREATION

After our tragic attempt to secure superpowers in Genesis 3, the Creator again takes center stage.[22] He does not come as a powerbroker with tools of mass destruction to eradicate the wrong, but he stoops down to work within our weakness, introducing a technology emerging from death itself: clothing. Skins divinely crafted from the death of a creature (Genesis 3:21) become reminders of lifeblood given for the sake of another. They provide a cover for now, until the long-term solution was revealed in the "fullness of time" through the death and resurrection of Jesus Christ—the firstborn of a renewed creation in which God himself will again dwell with us. Our true Genesis nakedness is not simply to be without clothes, it is to be without God.[23]

God is always faithful to his Word for creation. As sure as rain falls down and rainbows point up, God's creative word going out into chaos will not return without accomplishing its purpose (Isaiah 55:11). Because God is faithful, chaos will not rule. God assures us of this in Revelation 21:1-4, which is meant to be read in the context of Genesis 1:1-2.

> Then I saw "a new heaven and a new earth," for the first heaven and the first earth had passed away, and there was no longer any sea. I saw the Holy City, the new Jerusalem, coming down out of heaven from God, prepared as a bride beautifully dressed for her husband. And I heard a loud voice from the throne saying, "Look! God's dwelling place is now among the people, and he will dwell with them. They will be his people, and God himself will be with them and be their God. 'He will wipe every tear from their eyes. There will be no more death' or mourning or crying or pain, for the old order of things has passed away." (Revelation 21:1-4)

> In the beginning God created the heavens and the earth. Now the earth was formless and empty, darkness was over the surface of the deep, and the Spirit of God was hovering over the waters. (Genesis 1:1-2)

In biblical language "the waters" represent chaos, brokenness, and emptiness. When Revelation declares "no more sea," it is a restatement of Genesis 1; it is the same creating and re-creating Word of the Lord. The "waters" never had a chance of winning—not in a thousand years or in eternity. As extensive as the chaotic flood water that destroyed the earth was, the covenant promise fulfilled in Jesus Christ is even more extensive. Christ died for humans, for aphids, for zinc, and for everything in between. For God so loved the "cosmos" (not just humanity) that he gave his only son (John 3:16). Nothing in creation escapes the ravages of sin, but neither is anything in creation outside the bounds of the grace of God. God's love is unbounded and prodigal. The implication for engineering is that Christ cares about contaminated aquifers, jammed freeways, unventilated workplaces, and disfigured mine sites. God loves water, roads, warehouses, and mineral fields, and he also loves us, in both our brokenness and our potential.

God, through Christ, is now reconciling the world to himself. The demand for reconciliation has never been greater. As image bearers, we imitate him. And there is more work to do than there are available hands. We should read the second Great Commission literally.

> Then Jesus came to them and said, "All authority in heaven and on earth has been given to me. Therefore go and make disciples of all nations, baptizing them in the name of the Father and of the Son and of the Holy Spirit, and teaching them to obey everything I have commanded you. And surely I am with you always, to the very end of the age." (Matthew 28:18-20)

When Christ says, "All authority is given unto me," he is making it very clear: he doesn't need help, we need the work. God extends a job offer we cannot refuse—inviting us to go into all the world, training everyone who is longing for meaningful work and teaching them to "obey everything that I have commanded you." By "everything" he means the biblical narrative, from beginning to end, which includes the Spirit-filled task of developing and caring for the creation. He is returning us to the first Great Commission, inviting us to rest in the garden by keeping it, caring for it, and making the world a place where all things flourish. Sin has distracted us from our original job and created a mess that requires some healing and cleanup. For some of us, responding to this great recommission may involve engineering, but we shouldn't think we'll ever finish the project because its scope is far beyond our capability. We engineer to offer the world concept drawings of the kingdom yet to come, to let the world know that the ground has been broken, that resurrection-life is underway, and that it will be worth waiting (and working) for.

In God's plan, your work will not require relocation. God makes the move. Scripture is clear. Christ is returning. He is coming to this very place to stay. Just as God cleansed the world with a flood, so his refining fire of judgment will cleanse the heavens and earth (2 Peter 3:5-7). Our covenant-keeping God is making all things new. We will not recognize this old house when he is done with it. But it will still be the same house.

If something looks like "new life" and bears fruit, it is because the Spirit of God through Christ is working in and through creation in ways we cannot comprehend or imagine (John 3:1-21). When we find ways to bring beauty, hope, or healing to a broken world through engineering, it is by grace. When we see others (Christian or non-Christian) bring beauty, hope, or healing to a broken world through engineering, it is by grace.

The biblical meaning of spirituality is to be Spirit-filled. It is as unfathomable as Genesis 1:2, when "the Spirit of God was hovering over the waters." Spirituality is not something we achieve or maintain; it is what God does when the Spirit works over the chaos. It is as ordinary as engineering craft and practice when the Creator God says, "See, I have chosen Bezalel son of Uri, the son of Hur, of the tribe of Judah, and I have filled him with the Spirit of God, with wisdom, with understanding, with knowledge and with all kinds of skills" (Exodus 31:2-3). It is as pervasive and every-day-ordinary as Romans 8 and 12.

> Those who live according to the flesh have their minds set on what the flesh desires; but those who live in accordance with the Spirit have their minds set on what the Spirit desires. The mind governed by the flesh is death, but the mind governed by the Spirit is life and peace. (Romans 8:5-6)

> So here's what I want you to do, God helping you: Take your everyday, ordinary life—your sleeping, eating, going-to-work, and walking-around life—and place it before God as an offering. Embracing what God does for you is the best thing you can do for him. Don't become so well-adjusted to your culture that you fit into it without even thinking. Instead, fix your attention on God. You'll be changed from the inside out. Readily recognize what he

wants from you, and quickly respond to it. Unlike the culture around you, always dragging you down to its level of immaturity, God brings the best out of you, develops well-formed maturity in you. (Romans 12:1-2 *The Message*)

From beginning to end, Scripture tells the old, old story of the triune God and his love for his world, even as this world sets itself on a course of self-destruction. The Bible does not give specific arguments in favor of technology or engineering, but its stories show by example who we are created to be and how we are to stand, work, play, and live in right relation to our Creator. Culture making, from music to art to agriculture to engineering to astronomy, is neither argued for nor prescribed; it is assumed in the cultural mandate. The great human commission has always been to enable all creation to flourish. Scripture's answer to *how* we do that comes not as a recipe or formula but as a practice that stays attentive to what God is doing in Christ, both now and in the future, forever relearning to walk in step with the "creator spirit, by whose aid, the world's foundations first were laid."[24] This Spirit is as present in the office on Monday as it has been from Bezalel to Pentecost. The seventh day of restoration rest and re-creation has no end.

The Scriptures have much more to say about how we ought to live, including within our work related to technology. It should be a lifelong study for each of us to hear what God has to teach us about our culture making. Lest we think that our choices related to technology somehow fall beyond God's reign or somehow have no religious dimensions, the next chapter addresses the myth that technology is neutral, laying out the reasons that our designing and working with technology must be informed by the biblical story.

CHAPTER THREE

FIELD RESPONSIBILITY

Tools aren't neutral; rather, they encourage us and shape us toward certain goals, and they often do so in hidden ways.[1]
RICHARD R. GALLADERTZ

One of the most challenging aspects when exploring the field of technology revolves around responsibility. The act of engineering can often appear as purely objective calculations, especially the way textbooks teach it, as if every problem has exactly one solution. Field work is much messier than lab work, where problems sometimes have no solution and often have many possible solutions. Where there are multiple solutions, we have choices to make, and this is where responsibility begins. In this chapter, Steve VanderLeest examines several myths—false narratives that make it seem that no one is responsible. He then explores some of the ideas behind responsible and discerning design. Get ready to dive into a bit of philosophical debate here, but also keep your eyes open for some practical tips at the end of the journey.

Robert Moses built racism into his bridges. Moses was the twentieth-century architect of New York City, hailed as the "master builder" by the news media. As the commissioner of city parks, he was largely independent of city politics, yet he oversaw budgets in the millions of dollars. He developed hundreds of parks, hundreds of miles of parkways, and a dozen bridges. However, there was also a darker side to his legacy. He used extraordinarily low overpasses along roads leading to beach parks, designed to admit cars but not buses. An interview with a coworker by a Moses biographer indicated that this was intentional.[2] The evidence suggests that Moses designed the overpass to segregate, purposely biasing bridge technology to admit one class of citizen but discourage another. At that time, the Black population depended primarily on buses for transportation, while much of the White population drove automobiles. Whether or not Moses consciously made design decisions toward this end, his bridges produced an adverse effect: segregation. The social impacts of technology help us see that our design choices are not neutral. Design decisions can have political, racial, and moral biases, which may not be obvious.

Figure 3.1. Overpass on the Saw Mill River Parkway

Society sometimes makes it seem that technological choices are neutral, that our designs are purely the result of objective, unbiased mathematical calculations. Yet this is rarely the case. Technology design problems rarely have a single solution. There are usually choices to make. Those choices imply responsibility. This chapter challenges the myth of neutrality, addressing several of the arguments that falsely absolve us of responsibility for our technical choices. We then will look at why presuming technology design to be a morally neutral activity could be dangerous, concluding with some tips for approaching technology with Christian discernment.

You've probably heard the phrase "It's not the technology itself but what you do with it that counts." That statement has a ring of common sense yet fails to recognize the nuanced ways technology shapes us. What one can do with any particular technological gadget is limited by its design. The design is not neutral, as if it would allow you to do anything and everything with equal ease by using the tool. Some goals are readily achieved with a particular tool, and others are almost impossible because the tool was not designed with that goal in mind. The good or evil that someone does while using a tool is their responsibility. However, the engineer who designs a tool to particularly aid in those acts also bears some responsibility.

Each technological device has built-in biases and preferences that make some activities easier and others more difficult. "Embedded in every tool is an ideological bias, a predisposition to construct the world as one thing rather than another, to value one thing over another, to amplify one sense or skill or attitude more loudly than another."[3] Take the hammer for example. It is biased toward pounding nails and biased away from turning screws. If you place a hammer in someone's hand, they will be more inclined to pound nails and less inclined to turn a screw. So it is with all of our tools—their design implicitly makes some actions possible, makes some easy, and others difficult or impossible.

The tendencies built into our tools subtly change the way we think, work, and interact. Ultimately, those tendencies are rooted in our identity, our culture, and our faith. Some of these inherent preferences are good and God-honoring. Some are not. As designers of technology, the fact that a hammer is biased toward pounding should not worry us. What probably should worry us is technology that is structurally biased to preclude users based on gender or race. I suspect it is not easy to catch this. I hope to become more aware of the habits that a device might subtly encourage in my own life, especially if those habits change my behavior or attitudes for the worse.

While we might think of a city as a geographical location or as a community, we can also think of it as a technology. Philosopher Nicholas Wolterstorff notes the interplay in our relationship with technology when he writes about cities. He finds that inhabitants shape the city, but they are also influenced by how the previous inhabitants shaped it before them.

> Of technological possibilities, it is mainly the actions that city dwellers find to be necessary and desirable that determine their indoor/outdoor environment. And since these actions differ to some extent from city to city, each city is an expression of, a causal consequence of, the lifestyles of its residents. . . . A city, like a work of art, is value and rationality embedded in sensory material.[4]

As we change technology, technology changes us. To illustrate, the automobile has radically altered the shape of modern cities. A car is not just a neutral way to get from point A to point B—it fundamentally shapes our environment and culture. Today's cities are built around roads and streets that prioritize the automobile, making it more difficult and dangerous to bike or to walk. As the car extends how far we can travel in a short amount of time, the locations where we live, work, shop, and worship become increasingly dispersed.[5] The world before the car was a different world than the world after. So it goes with buildings, as Churchill observed when anticipating rebuilding after the war: "We shape our buildings, and afterwards our buildings shape us."[6] So it goes with all technology. In the words of one author, "We shape our tools and thereafter our tools shape us."[7]

Another hint that technology carries responsibility arises when things go wrong. We read about the adverse effects of technology almost daily. For example, a modern lawnmower has a bar beneath the push handle that must be clasped to engage the cutting blades. This safety interlock was added after people lost fingers attempting to trim hedges by picking up the running lawnmower and seriously injuring themselves in the process.[8] The mower story is not an isolated event. Despite safety mechanisms, warning labels, regulations, and training, we sometimes misuse technical gadgets in ways that harm ourselves or others. More troubling, however, is that injuries are sometimes a consequence of the way the technology was designed.

The design of technology reflects certain preferences of the designer and other stakeholders. These tendencies are inherent to the design of a tool. Thus, designers must think about the proclivities of the tool that might produce better results than they dreamed or worse results than they feared. It can even go both ways. For example, the designers of social media sites like Facebook did not anticipate that their technology might be used during natural disasters as a way for people to let loved ones know that they are safe. On the other hand, these same technology designers did not anticipate the level of hate speech and propaganda that would infect their sites. It turns out, in retrospect, we can see that the ease of instant posting of messages enabled both outcomes. It did not guarantee them, but the technology did have a proclivity toward both uses. To their credit, many of the large social media sites have recognized the value of communicating during emergencies and enhanced their services to encourage such use. They have also recognized the power of social media in sharing information and have put measures in place to discourage abuses such as trolling (online harassment) or blatant misrepresentation of the truth.

The design of technology can never be entirely objective since it reflects the will, desires, and even the personality of the engineer. Technology is a form of expression as much as music or writing, embodying our hopes and dreams. We build ourselves into our technology— sometimes intentionally, sometimes unwittingly. Our deepest beliefs become part of our designs, reflecting our worldview. The goal of this chapter and the next is to help you to be more technologically discerning as a faithful Christian designer or user of technology. We'll begin by dispelling some myths that prevent people from understanding or accepting responsibility for the gadgets we create. The next section will identify dangers in assuming that technical designs are value-free. Next, we'll examine technological mediation, a philosophy helpful for thinking about technology more robustly. Finally, we'll offer some tips for designing and using technology discerningly.

MYTHS OF BLAMELESS TECHNOLOGY

This section challenges false notions that technology is blameless, that technology design is neutral, that we are absolved of responsibility for our technological products.

Myth 1: Tools are universally useful. Some claim that technology design is neutral because tools are universally useful for whatever task we undertake. "Technology is neutral in the sense that it merely provides options; it does not compel anyone to take up those options."[9]

However, an obvious bias is the fact that tools are made to accomplish the specific purpose for which we designed them. We tend to use hammers to pound nails because we designed them to pound well. Can you also use a hammer to turn a wood screw? Not really—and not nearly as well as a screwdriver could perform that task. The tool's form tends to imply its function, and people tend to use tools according to their implied function. Whenever someone says they are "choosing the right tool for the job," they are implicitly acknowledging that tools are biased toward certain types of jobs.

It is also true that tools can fulfill more than one function, even in ways the designer did not imagine. Fuel-laden jet planes that terrorists used as missiles to bring down the Twin Towers of New York City on September 11, 2001, are one infamous example. Unplanned uses of technology can also be unexpectedly positive. In 1968, Spencer Silver, a scientist at 3M, was trying to find better adhesives. He came across one formulation that was not very strong but exhibited a peculiar ability to stick again and again. Although he circulated the idea throughout the company, no one appreciated its value. Then Art Fry realized Silver's idea might help him. Art was a member of a church choir but was frustrated that he would mark songs in the hymn book, but his little paper bookmarks often fell out. He needed something that would stick lightly to the page and could be removed later without damage. Thus, the Post-It note was born.[10]

The myth is that a variety of possible uses for one technology makes it neutral. The reality is that multiple uses indicate that there are complex biases enabling certain uses and discouraging others. Even so, we cannot hold aircraft designers responsible for the unanticipated September 11 hijacking attacks, and in general, we cannot hold designers at fault for unexpected evil uses of their technology. At least not the first time. However, once a harmful use is recognized, those that develop and sell the technology now have some responsibility to ensure that this injury is avoided. Avoiding the harm may be through legislation, such as reducing drunk driving by passing laws to regulate who can drive. Avoiding the harm may be through revising the technology itself, such as reinforcing the cockpit door of aircraft to impede terrorists.

Myth 2: Passive tools cannot be held responsible. Some claim that technology design is neutral because technology cannot make a moral choice—only humans can do so. A student in my class made this argument, claiming that computers and all technology are inert objects that are therefore neutral. Indeed, technological objects cannot make choices. They have no will or volition, no moral agency. A hammer cannot decide for itself whether to strike the head of a nail in carpentry or to strike the head of a person in murder. This decision is made by the person wielding the hammer. Under this theory, the designer of the hammer cannot be held responsible for murders perpetrated with the device they designed, since the tool is entirely neutral. Only the user makes ethical choices, and only the user is responsible for those choices.

However, while technological gadgets have no moral agency, they bear the biases and embed the values of the designer, because humans who design technology cannot help but impart values into their designs. Our moral agency cannot be cleanly separated from the objects that we design; technology always bears our fingerprints. Stated another way, technology such as a bridge or automobile cannot act on its own, but it nevertheless is value-laden, meaning that it embodies some of the values of the designer as tendencies and preferences. The technology's built-in functionality suggests and sometimes imposes constraints on the moral acts of people.

> **VALUE-LADEN** indicates something has embedded values, or something whose function, shape, or character was influenced by a set of values or a particular opinion.

Let's look at two examples. A surgical instrument designed to perform late-term abortions is not neutral. It was designed for a purpose that many Christians find abhorrent. Although that instrument could be used for other purposes, the designer should take some responsibility for enabling the act of the abortionist. Another example is a knee replacement implant. It was designed to replace natural joints that have worn out, thereby relieving significant pain and discomfort for the patient. It is hard to imagine using a knee replacement implant for purposes other than replacing a knee. Thus, although the surgeon can be thanked for the good work of performing the surgery, the engineer can also be thanked for the good work of designing the implant device. That is, both the user of the technology and the designer of the technology can and should take responsibility for good or ill effects.

Myth 3: The end justifies the means. Some claim that technology design is neutral because moral acts can only be judged by final consequences—the end justifies the means. The philosophy that underlies this approach is called *technological instrumentalism*. It views technology as simply a tool or instrument that we bend to our will. Any collateral damage must be judged in light of the good achieved.

The benefits do not always outweigh the damage caused by the means. However, when the goals for a technology do not warrant the side effects produced, those pursuing the goal may inappropriately rationalize that the detrimental effects are acceptable. As technology designers or users, we may too easily justify our ends, especially when the end is to our personal benefit. Cheating on a test because it produces a good grade does not justify cheating. On the other hand, studying hard is justified by the end goal of getting a good grade. When it comes to justifying technology choices, discerning where to draw the line may be best done by a wise team of peer reviewers who do not have a personal interest in the outcome.

Myth 4: Technological determinism. Some claim that we have no responsibility for the technology we develop by appealing to *technological determinism*, a philosophy that sees technology as driving history and social changes, envisioning an autonomous, inevitable impetus that results in specific technologies. A deterministic view can lead one to believe that there are no choices, that technological development is a foregone conclusion in which even the designer has no alternatives. If engineers have no choice and no influence over the technology design, then they have no responsibility either.

But technological determinism is not true. While technology does have a strong influence on our cultural identity and social structures, it does not leave us without our influence. Technology is not an abstract, outside force that is out of the control of humans. It only appears that way because it is influenced by more than one human, by more than the engineer. It is influenced by the sociocultural environment in which it is imagined and implemented.

Technology is a rich tapestry woven from threads of science, economics, government, and societal structures like families and churches. Technology influences history but is also influenced by history. It is the product of human choices and biases. Some influences are beyond the reach of the engineers designing the technology, but engineers can and must make choices. In these choices, we exercise freedom and responsibility.

The evidence for choice in design can be found in the wide variety of technologies designed for the same purpose. Consider toilets; more specifically, consider how to flush them. There is more than one design for this mechanism. Toilets in the United States have a flush lever on the side, while European toilets have two push buttons on top.

Design choices are always constrained. You can make a car safer by increasing the amount of sturdy metal framework around its passengers, but this adds weight so the fuel efficiency drops. There are always tradeoffs in design, which may give the appearance that some choices are predetermined. However, selecting wise alternatives is part of what makes engineering interesting. It is why we need creative people to design technology. Ultimately, because technology is not deterministic, engineers have responsibility for the choices they make.

Myth 5: They made me do it. One more way to avoid responsibility for technology choices is by insisting that we were coerced. Innocence is claimed because the boss told us to do it. Most engineers on the job have experienced this feeling—that they had no choice in which products to design, that even the detailed design decisions were dictated from above. There is some truth to this feeling.

Due to the communal nature of design, rarely is one person responsible for all design decisions. In a large company, many departments will have an influence on design decisions, including management, marketing, finance, and more. Some decisions will seem to be made based on internal politics more than on technical merits. In such a situation, I could rationalize that the decision is no longer my responsibility. However, if the decision is going in a way that seems wrong, the engineers on the team may need to have the courage to speak up. When the consequences of a bad decision may be dire, silent submission is tacit consent that makes the entire team culpable.

The accountability for design choices extends even beyond the company developing the product to suppliers and customers, and thus to society as a whole, which creates a market demand for the product. We will see an example of this complex interplay of design decision drivers later in chapter seven, where we explore the historical choice between electric and gas-powered vehicles.

THE DANGERS OF TECHNOLOGY DESIGNED IRRESPONSIBLY

The belief that technology development is completely objective can obscure the big picture, persuading us to wrongly believe that one size fits all, blinding us to alternative approaches so that we fail to recognize important consequences of our decisions. We see the trees of our individual decisions in designing technology but miss the forest of the cumulative responsibility for the impact of that technology. We would never intentionally contribute to the design of a racist bridge. However, without careful discernment, we might do so accidentally by overlooking customers that use a variety of transportation modes. Bias can sneak into design. Consider two real-world examples: calculating machines for the Nazis and the implicitly racist Shirley card.

In the 1930s, International Business Machines (IBM) was designing punch card systems to enhance the efficiency of train schedules. They excelled at designing such equipment, ensuring that these machines could quickly and accurately compute schedules. The engineers focused on customizing their general-purpose calculating devices to compute the schedule for a complex network of trains. They likely thought efficiency was the end of their responsibility. They did not consider that designing this system raised broader moral questions—even though their customer was Nazi Germany.

Hitler's Third Reich was using the calculating machines to improve the effectiveness of their program to exterminate the Jews. Furthermore, according to at least one published report, IBM knew the end purposes of their customer, yet continued to work closely with them until the US entered the war.[11] While we properly place the blame for wartime atrocities on the commanders who directed such horrific acts and on the soldiers who knowingly carried them out, we can also find some culpability for the people that aided them. The engineers and managers at IBM were at least negligent and perhaps maliciously greedy if they knowingly supplied equipment to aid such evil acts.

A second example comes from the technology of photography. Sarah Lewis, an assistant professor at Harvard University, has studied imaging technology, finding it contains some troubling biases—biases that have even managed to migrate across generations of the technology.

> Photography is not just a system of calibrating light, but a technology of subjective decisions. Light skin became the chemical baseline for film technology, fulfilling the needs of its target dominant market. For example, developing color-film technology initially required what was called a Shirley card. When you sent off your film to get developed, lab technicians would use the image of a white woman with brown hair named Shirley as the measuring stick against which they calibrated the colors. Quality control meant ensuring that Shirley's face looked good. It has translated into the color-balancing of digital technology.[12]

Imaging bias still exists today. Modern digital imaging technology carries the remnants of this bias, displaying darker skin poorly by default. The racial bias toward a certain skin color started in early film cameras, but it did not stop there. It jumped the gap to digital and malignantly embedded algorithms in modern facial recognition technology.

Every technology designer needs to consider biases and tendencies of all types: social, ethical, environmental, racial, gender, cultural, philosophical, and more. Despite our best efforts, our designs will likely fall short of that goal; and when they fall far short, we may need to seek forgiveness from those we have unintentionally hurt through our technology.

TAKING RESPONSIBILITY FOR
TECHNOLOGY DESIGN

In 1978, a beautiful hotel was completed in Kansas City, Missouri, featuring a visually stunning atrium with suspended walkways between upper floors extending over the lobby. Engineers achieved the open look of the walkways through the clever use of suspension rods vertically extended through walkways at multiple levels. Not long after that grand opening, the hotel hosted a large event, with guests crowding all areas of the atrium, including the walkways. The guests mingled over appetizers and drinks while listening to music, oblivious to a critical flaw in the walkway construction, unaware that many of them would not survive that evening. At 7:05 p.m., around two dozen attendees on the fourth-floor walkway felt a gut-wrenching vertical shift as the walkway dropped several inches, followed by what witnesses described as a horrific scene. The supports of the fourth-floor walkway then completely gave way, plummeting down to the second-floor walkway, which also gave way under the rain of concrete, crashing to the lobby floor below. Rescue squads were able to save many of the sixteen hundred people present that night, but over one hundred died in the tragedy.

This deadly structural failure at the hotel was traced back to a poor design change. A judge "found the structural engineers for the Hyatt Regency Hotel guilty of 'gross negligence' in the 1981 collapse of two suspended walkways in the hotel lobby that killed 114 people."[13] Figure 3.3 illustrates the lethal change. The original design, on the left, specified a single rod to extend from the ceiling through three levels of floors, with a threaded nut beneath each floor holding the weight of a single floor. In this original design, the threaded nuts specified for the project were sufficiently strong to hold one floor. The revised design, on the right, shows that three shorter rods were substituted in place of one long rod, putting the weight of two floors on a threaded nut that was only capable of bearing the weight of one. The charge of negligence indicated that the engineers approving the change did not exercise sufficient care in analyzing the new design to ensure it was still safe.

Figure 3.2. Hyatt Regency walkway collapse

The tragic fatality count from the Hyatt Regency walkway collapse would not be surpassed until thousands died in the fall of the Twin Towers of the World Trade Center in New York on the 11th of September in 2001. Terrorists made malicious use of technology, hijacking jetliners full of fuel to crash them into the towers so that the resulting fires would weaken the structure. Multiple investigations afterward pointed to the fact that the extreme heat of the burning fuel weakened the steel skeleton of the building, causing it to collapse, killing thousands. In this case, although the mechanism of failure was identified, moral accountability was generally not assigned to the designers of the building. Amazingly, the buildings withstood the initial impact of the jet aircraft, and it was only the added element of burning jet fuel that pushed them past their limits. Thus, the building architects were not held responsible and neither were the aircraft designers. At the same time, the malicious misuse of technology by the terrorists was an obvious moral wrong.

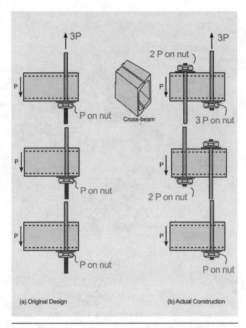

Figure 3.3. Fatal design change in Hyatt Regency walkway

We get a glimpse of our responsibility for the technology that we develop when flaws are discovered in a design. A purchased tool that does not function as advertised can be returned for a refund, whether the dysfunction is caused by a defect in the materials or the design. A tool with a design flaw that causes injury or damage can result in legal action, demonstrating that civil society holds designers responsible for their designs and responsible for serious flaws in those designs. The law considers it a malicious act if designers knowingly develop flawed technology or when they discover a flaw later and do not correct it.

If designers miss a flaw that should have been eliminated by normal design practices, then they are considered guilty of negligence. In the early 1970s, engineers located the gas tank of the Ford Pinto toward the rear of the vehicle, a position known to be dangerous in a rear-end collision. Several fires during accidents gave direct evidence of the flaw, and litigation over this problem resulted in the largest punitive damages ever awarded at the time.

Designers face some moral culpability for their flawed designs whether the flaws are due to ethical choices of commission or omission. These choices can be identified later, to the point of assigning responsibility for a flaw. Scripture points us toward responsibility for the designer as well: "When you build a new house, make a parapet around your roof so that you may not bring the guilt of bloodshed on your house if someone falls from the roof" (Deuteronomy 22:8). God leaves no room for excuses here. "But he shouldn't have been so close to the edge. He wasn't watching where he was walking!" God is having none of that, holding the builders responsible for the safety of their guests. This passage also implicitly assigns some responsibility to the guest. A parapet is a low wall, not a complete barrier, and this is sufficient to protect the guest. Guests could still fall by doing something reckless while on the roof, but now it would be their fault and the homeowner would be guiltless. The implications of this verse should be striking for those of us designing technology. If God expected the builder of a home to anticipate safety issues for guests, then it seems that God would also expect technology designers to anticipate safety issues for our guests, that is, the customers, users, and stakeholders of our technological products.

Responsibility for technology is shared by a broad group of stakeholders in part because it is interwoven within social structures. Consider technology that is intertwined with government organizations. In this case, technology might not seem like a problem by itself, but in the hands of bureaucratic agencies believing it will make them more efficient, one could imagine the combination resulting in negative impacts. For example, a state unemployment agency might put their benefits application form online, creating obstacles for the unemployed without a home computer—yet these are likely the poorest of applicants that are most in need of help. Technology carries additional moral weight as soon as it is combined with other social or cultural elements. All those who had a hand in the invention, design, and deployment of a technology have some responsibility for its impact, including scientists, engineers, machinists, managers, sales staff, and more. The full weight of the responsibility for the impact of technology is not on the shoulders of the designer alone, it is shared by all those contributing to the shaping of the technology as it is interwoven into our societal structures.

Even if responsibility is shared, it is still weighty. Shirking that responsibility can be downright dangerous because technology will always carry the designer's values (and perhaps also the values of other stakeholders). The peril may be unintended, but it still has consequences.

RESPONSIBILITY EVEN FOR UNINTENDED CONSEQUENCES

In criminal law, intent to commit a crime is part of the equation for assigning guilt. However, even if not intentional, if defendants could reasonably foresee their actions would cause harm, they can be found guilty of negligence. Likewise, with the design of technology, we have a responsibility to anticipate what might happen. If harm can be reasonably foreseen, we have a responsibility to prevent it.

Presuming technology comes without responsibility can lead us to think that one size fits all. Without intending to discriminate, we can unintentionally design products that only fit ourselves. For example, we know that car drivers are not all one size, yet it is easy to design an automobile interior for someone of roughly our own height, resulting in taller occupants bumping their heads on the ceiling and shorter occupants having trouble seeing over the dashboard.

Likewise, building visitors are not all the same. For many years, architects in the US and elsewhere designed buildings with exterior entry steps and interior stairways between floors without much thought about making the buildings accessible to all. If they did consider accessibility, cost concerns often eliminated those options. These buildings were not neutral—they were designed with a bias against accessibility to those that could not easily traverse steps. Humans are diverse. Flexible designs that fit the needs of the whole community may cost more than those that cater to a particular height, gender, or race. In some cases, the community has enacted regulations to ensure the needs of all are met, such as the Americans with Disabilities Act ensuring fair access to buildings.

Presuming that technology comes without responsibility can blind us to alternative uses of a product. We design a device to perform well for the people and situations we have in mind. But different people in different situations sometimes use a device in unexpected ways, causing it to fail, or worse, resulting in injury or property damage. Part of an engineer's professional responsibility is to anticipate alternative uses and abuses of a product. Unanticipated use leads to unintended consequences. These results, in turn, can lead to claims of negligence against engineers who should have thought of this possibility and taken measures to prevent it.

Of course, engineers are not omniscient and cannot think of everything, but they should be reasonably imaginative to be able to identify likely alternatives. The history of similar products and market research on envisioned products can improve our predictions. Nevertheless, it can be difficult to know how deep to dive into these predictions. Some of the best judges of due diligence should be our fellow engineers. We ought to hold each other to a high standard, but also recognize what is reasonable and feasible to foresee. A few examples can help us think about this further.

When Pokémon GO was released as a smartphone app, the developers probably should have realized that players of this augmented reality mobile game would become completely immersed while attempting to catch Pokémon. A few users became so immersed that they did not notice they were walking into traffic or other dangerous situations. The developers could have anticipated this and tuned the software to avoid mapping virtual game objectives into treacherous locations in the real world.

Likewise, a little more research could have prevented harm that initially resulted from automobile airbags. The intent of this new feature was certainly noble: to prevent injury during a crash. An accelerometer detected the telltale sign of a collision: rapid deceleration. This signal then triggered the explosive expansion of an airbag in front of the passenger, expanding so rapidly that it cushioned the body during the milliseconds before the occupant struck the steering wheel or dashboard, preventing massive injuries. The foreseeable problem was that, while the system worked well for an average-sized adult, it could harm a child. After a number of children suffered serious injuries from airbags, laws were enacted that prohibited children from sitting in the front seat. Could this problem have been anticipated? Yes. Testing for a wider range of occupant sizes would have revealed the issue, so the engineers had some responsibility for missing a crucial implication of the design.

Some unanticipated consequences are so far-fetched that it is difficult to imagine one could have thought of them ahead of time and done something to prevent abuse or harm. For example, another side effect of Pokémon GO was abuse by criminals to lure unsuspecting players to secluded areas where they could be robbed.[14] Since the game allowed players to form teams and communicate their physical locations, thieves abused the technology to set up their victims. Broadly speaking, in cases like this, the abuser of the technology is responsible, not the engineer.

Although the Pokémon GO designers are not directly accountable for virtual highway robbers, having learned of this problem, they should update the game to provide better security protection for players. The Old Testament law speaks to this idea of initial and subsequent responsibility. "If a bull gores a man or woman to death, . . . the owner of the bull will not be held responsible" (Exodus 21:28). The cattle owner is not responsible, at least initially. Yet the passage goes on: "If, however, the bull has had the habit of goring and the owner has been warned but has not kept it penned up and it kills a man or woman, the bull is to be stoned and its owner also is to be put to death" (v. 29). Once we are warned of the harm caused by any of our possessions—including our technology—then it seems God now holds us accountable. This idea that engineers can be held accountable for the technological gadgets they design can also be found in laws regarding product liability, which generally hold the designer and manufacturer of a device to be responsible if the device causes harm.

Accountability for a design that has caused harm could play out in several directions. We may choose to revise the design to prevent the harm we now know the first version could cause. We might choose to seek forgiveness from those that were harmed. We might seek to restore the well-being of the harmed users.

Some harmful consequences can be anticipated but not easily prevented with technical solutions. For example, a hammer is a simple tool designed for pounding nails. However, it can become a weapon in the hands of an angry murderer. It is hard to imagine how we could design the hammer to pound nails effectively and yet prevent its use as a weapon, so the engineer's responsibility for preventing such abuse is limited.

Not all the results of assuming technology to be neutral result in harmful unanticipated consequences. We also miss out on potential unforeseen benefits. Sometimes users find delightful new uses for a technology that the designer did not intend, such as Facebook users notifying friends and family that they are okay after a disaster. This was not an intended use for status updates but seemed so useful and important that Facebook formalized this use and went on to provide quick ways to make such posts.

Another example of fortuitous unanticipated consequences is the use of cellphones in African countries where the installation of telephone landlines was proceeding too slowly. Using mobile technologies not only allowed them to skip physically wired technology, but it also started replacing certain financial functions. Kenyans, wary of the stability of their local banks, started using prepaid phone accounts as a way to more safely save money and transfer funds to others. They also used phones to look up real-time financial data on commodities to ensure a fair price for their crops. Technology users who creatively find new uses for a gadget are, in a way, doing engineering and technology development themselves. This is not surprising since all humans bear God's image and have creative abilities.

Spider-Man's uncle tells him, "With great power comes great responsibility." We've made the case that technology design comes with responsibility and presuming otherwise can produce undesirable consequences. Knowing this, how can we as engineers and scientists better work with technology? One philosophical approach to technology that might help is the idea of *technological mediation*, an approach to designing with discernment because it explicitly recognizes the biases we build into our gadgets.

TECHNOLOGICAL MEDIATION

Think about holding a hammer. The form of this handy tool implies its function—you want to pound things with it. The hammer makes you perceive things around you as nails that need pounding. The instrument influences the wielder so that your view of the world is biased toward the uses the instrument implies. The heft of the hammer's head along with the long handle providing leverage contributes to its ability to amplify the power of your arm. Almost all technology has this innate ability to amplify human ability. Marshall McLuhan said that "all technologies are extensions of our physical and nervous systems to increase power and speed."[15]

A similar idea appears in the philosophy of Don Ihde: all technology-mediated relations have a basic amplification-reduction structure. Like a magnifying glass, certain features are amplified, but the field of vision is narrowed. "Ihde argues against any view of technology as neutral . . . a perspective that relies on emphasizing amplification while ignoring reduction. At the same time, he rejects the idea that technology is a 'Frankenstein phenomenon' opposed to the human—a view that emphasizes only reduction while ignoring amplification."[16]

Ihde's point is significant. If users only notice how technology amplifies, they might mistakenly think it is neutral, giving them the power to do anything with it they choose. The choice appears to be purely their own. However, each amplification also comes with a reduction. Restrictions mean the tool is not neutral—it enables some choices but disables others. When you peer into a telescope, your peripheral vision is blocked, limiting the overall view and context. Holding a hammer makes it easy to pound nails but not to put eggs in the fridge.

I have seen this amplification-reduction effect at work in my own choices about technology. Perhaps you have too. Consider choosing between an automobile or a bicycle for a short distance journey. The bike provides the opportunity to exercise and can take paths the car cannot. The car will keep you dry in a rainstorm but may take more time to park upon reaching your destination. Those same barriers that the car provides to keep you dry also isolate you from your surroundings more than a bike. Technology amplifies some abilities while narrowing others.

The amplification-reduction model does not simply capture a pair of opposites. The reality of our technological society creates a more complex landscape between these two ideas. Ihde suggested four types of mediating relationships between humans, technology, and the world, which philosopher Peter-Paul Verbeek expanded to seven:[17]

▶ *Embodiment* relations combine humans and technology to relate to the world, often using a "through" preposition. We speak through a phone to another person. We look through a microscope to examine tiny bacteria.

▶ *Hermeneutic* relations put technology between the world and humans, often with an "at" or "to" preposition. We look at a thermometer to determine the temperature of the turkey baking in our oven for dinner. We listen to the beep of a metal detector to realize we found a valuable ring buried in the beach sand below our feet.

▸ *Alterity* relations describe how humans and technology interact with each other with the world in the background. These are often described with directive verbs such as "command" or "select." We direct the ATM to dispense money. We select puree on the kitchen blender.

▸ *Background* relations combine technology and the world almost transparently so that the technology is part of the experienced environment, often using adjectives regarding our feelings or comfort. I feel cool because of the air conditioning in my home. I can see the room clearly because of the good lighting.

▸ In *Cyborg* relations, the technology is intimately combined with humans so that they jointly interact with the world, such as cochlear implants.

▸ *Immersion* relations are similar to background relations, but here the technology provides an interactive context for humans, such as home assistants that recognize your voice and adjust the home lighting and temperature settings.

▸ *Augmentation* relations are a combination of background and hermeneutic relations, where the technology mediates between the human and the world but also alters our perception of the world, such as Augmented Reality (AR) vision provided by modern smartphones.

In thinking about the mediating relationships between humans, technology, and the world, Christians also recognize another vital participant: God. God the Father is the Creator who endowed his creation with rich possibilities, including the possibility for technology. The creation glorifies the Creator and thus technology should be designed with that relationship in mind. God the Son is the Savior who redeems us and all of creation. Creation, including technology, groans under the weight of sin. The mediation of technology can also bind or restrict us. Recognizing those potential restrictions and the value-ladenness of technology, we can aim for technology designs that help humans be redemptive agents in a fallen world, with the help of God the Holy Spirit as our counselor. Technology should be guided by the third person of the Trinity, helping us discern technology's proper place and purpose. We close this chapter with a few practical tips for such discerning.

TIPS FOR DISCERNING DESIGN

The designer is evident in the design. We build our hopes and dreams into our designs, as well as our flaws and faults. We should thus design with discernment. Someone with a discerning eye has good taste and good judgment. A discerning member of the audience at a symphony performance recognizes the skill of the musician in a superb musical performance. A discerning wine taster can distinguish a classic vintage over a knockoff store brand. Christians are called to spiritual discernment that goes beyond judgments of quality or excellence, though these are important. Spiritual discernment includes the ability to make choices recognizing and pursuing goodness in the world around us while recognizing and fleeing from sin.

Christian engineers have the high calling to use such discernment as they design technology. God invites us to become his redemptive agents, designing for shalom. Up to this point, this chapter has made the case that technology carries responsibility. This closing section will suggest some guidelines for taking that responsibility seriously, designing technology with discernment by exploring the user's perspective, reverse engineering, and holistic thinking. The next chapter will go further, providing a systematic framework for normative design.

First, discernment can be gained by walking a mile in the shoes of the user. To be discerning, we should think like a user. Doing so widens perspective so that we better anticipate unintended uses of our product. It is easy for designers to be somewhat blinded to unintended uses because the intended use is so clear in our minds. In designing footwear, it might not occur to us that someone would use the hard heel as an improvised hammer to pound a nail into the wall when hanging a picture in their new apartment.

Getting out of the rut of your original thinking can be challenging. Consider how the product might appear to someone seeing it for the first time. Does it resemble anything? Consider the individual elements that make up your product. How might someone use each element? Consider the whole product, but imagine a different goal, perhaps even one that is silly, to stoke your creative process.

Thinking like a user not only helps identify unintended uses but also improves our approach to the intended use. We might learn that the child car seat we designed is not easy to use for a parent with relatively short arms. We might learn that the signage labeling a dangerous element of our machine is not easily read by colorblind users. We might learn that the handle on our new tool is not comfortable after extended use by someone with large hands or is awkward for left-handers.

We may not always be able to think like a user. If you have trouble imagining alternative uses for your product, you could use early prototypes of the product with a focus group of users. You might even choose to keep the intended use secret at first, to get the group's unbiased first impressions. Keep safety in mind at this early stage. Even a prototype can be dangerous if not fully tested or if used in an unintended way.

A design approach that recognizes the importance of seeing the design from the user's perspective is the Agile methodology for software development. One of the tenets of Agile development is to produce working software after each iteration of the design (called a sprint) with the express purpose of allowing users to give the software a test drive. This continuous build-test cycle directs coding activity toward the best value for the customer, allowing the user to provide input throughout the design phases. This type of beta-testing is not limited to software, some form of early prototype testing can be done in all areas of technology.

Second, we can improve our design discernment skills through an exercise of reverse engineering. In this case, our goal is not only to understand the intended function but also to recognize inherent bias. By working backward from the finished product to identify the intended usage, we beef up our discernment skills and enhance our ability to recognize that any tool can serve multiple purposes. Using such an analysis exercises our ability to see designs as complex, multifaceted tools. Start with an unfamiliar device, perhaps one that you found or came across in a garage sale. It is best if you do not have any literature or labels to indicate the intended purpose of the product. Look at it from every angle. What aspects of its form imply its function? You might even consider taking it apart—perhaps the guts of the machine will spill its secrets. You can also practice reverse engineering on products for which you know the intended purpose. In this case, brainstorm how to use the tool in alternative ways. What additional goals could you accomplish with this tool, perhaps even humorous ones? Such creative thinking broadens your perspective so that you can recognize the intricate and adaptable nature of the tools you design.

Another exercise in discernment is familiar to most of us: comparison shopping. Ranking several products against desired features is a way to evaluate the intended purpose of a product and judge it against your own design criteria. For example, I recently considered products to heat water in my home. I ranked traditional storage tank water heaters, tankless water heaters, and heat pump (hybrid) water heaters. My criteria included the ability to provide hot water for simultaneous uses (like running the dishwasher at the same time someone is in the shower), efficiency, initial cost, and expected lifetime. By identifying my goals for the purchase, I was setting up my design requirements and specifications. By rating the products I was considering against my design criteria, I was discerning how well the product met my intended usage. The selection criteria I listed give some good hints about my values. You will likely do something similar when shopping for your next mobile device, next car, next pair of shoes, or almost any important purchase.

A **DECISION MATRIX** is a table with rows that represent different design alternatives/options and columns that represent decision criteria, such as low cost, high reliability, good fuel efficiency, and so on. The entries in the table are then filled in with numerical scores to rate each design alternative against each criterion. The total of each row then provides an overall score for how well the design option meets the criteria. Sometimes the criteria are weighted differently and the scores are scaled accordingly.

Third, we can design with discernment through holistic and contextual thinking, thinking about the broader impacts of the technology: economic, political, aesthetic, ethical, environmental, and more. Technology design is not done in a vacuum, it is embedded in the complex warp and woof of life. Engineering is only one thread of cultural activity woven together with other aspects of our culture, society, and world. Once introduced into the market, technology must function holistically within that complex world. Therefore, our designing must not only consider technical aspects like the appropriate weight of the product or the proper resolution for the user display, but it must also holistically consider the beauty of the design, the recyclability of the materials, the impact of the design on community and family life, and much more.

The design norms we will introduce in the next chapter offer a framework that can help you think broadly and contextually about your design choices. Imagine what would happen if our product decision matrix had design criteria that included not only cost, speed, weight, and other technical goals, but also included justice, caring, and stewardship. We might then see the whole cultural forest and not just individual trees.

CONCLUSION

This chapter covered one of the more challenging concepts for students and professionals in engineering: the non-neutrality of technology design. Even if you are not convinced, hopefully the provided examples and examinations of myths have prompted you to think further about whether some of our engineering might be more subjective than we realize.

Besides using design norms in the next chapter to stir our thinking and help us see the context of our work, engineers should expect and welcome input from experts in other areas. We should have a lifelong goal to become true renaissance engineers, familiar not only with the technical disciplines central to the profession but also with the humanities, the arts, and the social sciences—in sum, the broad sociocultural symphony in which technology blends its voice. To contribute best to our society, to serve our neighbor, and to glorify God, it is not enough to do the math and the science well. Technical work must be done with discernment in service to greater and more noble ends.

CHAPTER FOUR

FAITHFUL DESIGN GUIDES

The pursuit of these three things together—holism, proper design principles,
and appropriate design specifications—is what we call sufficient design.[1]
STEPHEN MONSMA

Engineers love tools. At its core, technology is fundamentally a tool, an instrument that we design to achieve some purpose. While you might expect tools like a magnifying glass or a butterfly net for a biological field trip, what kinds of instruments do you need for designing tech? Earlier generation engineers used tools such as a slide rule or a trusty calculator. Modern engineers might use a high-end laptop or perhaps even Virtual Reality goggles. Engineers trying to live out their faith can also use some instruments. In this next chapter, Steve VanderLeest describes some tools that Christians can use for tech design: normative principles for rightly creating and evaluating technology. We will put these norms to work in later chapters, so think of this chapter as setting up base camp for further field trips.

Madison loved designing web sites. She considered herself a multi-talented web developer, drawing on the diverse skills of user experience, graphic design, web server programming protocols, and more. The websites she designed for her clients were beautiful, functional, intuitive, and cleverly crafted to allow the customer's business to quickly scale up without losing performance. Madison loved the feedback she got from her customers and even the kudos from her customers'

customers who were the ultimate users of her work, those that really appreciated her art. Madison loved the fact that she could help people in her job using her God-given talents. She was good at developing high-performance backend web services. She used the most current web development tools and languages and prided herself in the blinding speed with which she could churn out great web pages. Her pages were visually appealing and she regularly delivered the power of the web to her customers on time and on budget.

Then she got a feedback comment that undermined her tidy view of her professional work. The email read:

> Dear Webmaster: I wanted to use your website to purchase some items that I have heard would be quite useful to me. However, I am legally blind. Your site is arranged in a way my seeing friends tell me is quite beautiful from a visual perspective. However, I "see" your site by using a screen reader browser extension designed to "read" a website for those that are visually impaired by translating the page and reading it aloud. Web sites can still contain plenty of visuals and yet can be easy for me to use and appreciate as well. For example, images can be tagged with alternative text that is read out to me so that I get a description of the image. (The WC3 organization[2] has plenty of other tips like this.) I urge you to consider some revision of your designs to make them more accessible to would-be customers like me.

Madison was devastated by this email. She had been so focused on the customers as she had envisioned them that she had neglected to consider that not all customers had good vision. She was personally committed to God's call to particularly care for those less fortunate in society, and yet had not considered this segment of her audience of users. Madison had too quickly accepted the limitations of the technology and presumed that since the web was obviously designed for visual effects, they were what made a website effective. She had betrayed her own noble intention of helping people through her web development work. Madison had good intentions, but designing with good intentions does not avoid bad outcomes.

Similar examples could probably be cited by many designers. Our modern, technological culture often encourages reductionism—the assumption that a system is nothing more than the sum of its parts. Simply analyzing components can make us miss the big picture. Design must also be done with good attention to context and broader goals that are consistent with our faith.

Just as early sketches must move to shop drawings with more specific details about the product, engineers need additional guidance to discern the shape that faithful design will take. We can find such guidance both in Scripture, God's special revelation to us, and in nature, God's creational revelation to us. "It is by listening to the voice of God in the work of his hands that the farmer finds the way of agricultural wisdom."[3] As with the farmer, the "same is true in economics, politics, the arts, medicine, communications, and education—in every area of society. We learn how to take care of God's creation by familiarizing ourselves with the creational structures and living in tune with them, and we formalize that knowledge in a Christian worldview."[4]

This chapter outlines a framework to guide designing technology with discernment. Although there have been other frameworks proposed for this purpose, including those known as *value sensitive design* and *appropriate technology*, the approach that all three authors believe can be especially helpful for Christians working in technology is the approach of *design norms*. These norms are based on biblical principles as well as insights from creational revelation. The norms are a framework to help focus decision-making and base it on a Christian worldview, a means for your faith to infuse your engineering design. They give you a place to start that leads to lifelong discipleship and the advancing of God's kingdom within technical careers. The design norms focus on the content of our designs, with some application to personal behavior. This is in contrast to a virtue-based approach, discussed in the next chapter, which focuses on our individual behavior with some application to the content of our design decisions. The two are not mutually exclusive—how we behave in the workplace likely influences our design choices as well.

DESIGN NORMS

When designing a system according to a set of requirements, we must validate and verify. Validation is the process of ensuring we "did the *right* thing." We ensure that our set of requirements is complete—there are no missing specifications that are important to the success of the design. We check that the specifications make sense and are consistent. Once we know we have targeted the thing we should, then verification is the process of ensuring we "did the *thing* right." We ensure that the design correctly implements the specifications, checking this by analysis, test, and measurement. Design assurance through validation and verification is especially important for safety-critical designs, where a missing requirement or invalid implementation could cost lives. However, for truly good design that honors God—what the authors of *Responsible Technology* termed *sufficient design*—ultimately we must look beyond technical requirements and consider what broader aspirations we should pursue for all our technology development. This chapter explores these overarching aspirations, which we will call *norms*.

Norm may not be an everyday term. Short for normative principle, a norm is an authoritative standard or rule. Think of it as a guide to better behavior, a suggestion for what we ought to do. A set of norms for technology design can help steer us away from dangers such as reductionism. The design norms presented in this chapter can guide our design of technology. They are based on biblical principles as seen through the lens of the Christian philosopher Herman Dooyeweerd.

Born in Amsterdam, Dooyeweerd was a professor of law for many years but is most famous for his philosophy of the *cosmonomic idea*—a way to think about reality based on a Christian worldview. Although he didn't focus specifically on technology, Dooyeweerd's efforts to describe all of reality in categories that he called "modal aspects" provide a useful way to think about the world that can be adapted as norms for technology design. Dooyeweerd's modal aspects are listed below, along with a few descriptive words to get a better sense of what he meant by them. Dooyeweerd acknowledged that human rationality was finite and thus his list might not be complete. In fact, he imagined new aspects might be discovered in the future.

Table 4.1. Modal aspects of reality

Numeric	discrete quantity
Spatial	continuous extension
Kinematic	movement, flow
Physical	energy, matter, force, current
Biotic/Organic	life functions, organisms
Sensitive/Psychic	sense, feeling, emotion
Analytical	distinction, conceptualization, logical, categorizing, clarifying
Formative	shaping of history, culture, technology
Lingual	symbolizing meaning to enable communication
Social	social relationships, interactions, roles
Economic	frugal management, care and use of resources
Aesthetic	harmony, surprise, fun, play, enjoyment
Juridical	what is due: rights, responsibility
Ethical	self-giving (agape) love, generosity
Pistic	faith, vision, commitment, belief

Dooyeweerd claimed that each aspect had its own distinct character: none could be merged into or subsumed by another without losing something. Furthermore, starting with the analytical and moving to the pistic, he saw the last nine aspects as uniquely involving human interaction, describing us as individuals and collectively in relationships. The later aspects build on and emerge from earlier aspects. A given aspect cannot be reduced to another aspect. For example, the biotic aspect cannot be explained by physical laws. The aesthetic aspect cannot be explained by purely social or economic rules.

> **REDUCTIONISM** is a philosophy that assumes complex behavior can be explained simply as the sum of the parts, thus minimizing or completely dismissing the importance of the interactions between the parts and emergent behavior that may not be apparent from any individual element or combination of elements.

As a framework that helps us understand the nature of reality, these aspects help us think about how to design technology. A team of authors at the Calvin Center for Christian Scholarship described technology design norms based on modal aspects in their book *Responsible Technology*.[5] One of those authors, Egbert Schuurman, went on to write several more books that outlined technology design norms, providing a more fluid list that sometimes included norms such as effectiveness (tied to analytical and economic aspects), care and respect (tied to the ethical aspect), and service, trust, and faith (tied to the pistic aspect).[6] Ermer describes using design norms within a standard engineering design tool, the decision matrix.[7] The traditional matrix evaluates design alternatives on the basis of decision criteria such as cost, time to market, reliability, and so forth. However, the decision criteria should also include design norms, prompting a deeper consideration of the impact new technology can have on our society and our world. One of the other authors of this book (Derek) has also written about applying Dooyeweerd's aspects to norms specifically for digital technology.[8]

Dooyeweerd's first six modal aspects describe nonhuman reality, aspects for things like the mathematical operation of addition or the physical force of gravity. These first six aspects apply to all creation and all creatures and provide the context for human activity. Engineers and others working with technology tend to apply these aspects instinctively in their work. However, the last nine may not come as naturally when designing technology yet are just as important. Beginning with the analytical, they describe distinctly human ways of acting.

The rest of this chapter explores some details and examples of design norms based on Dooyeweerd's way of looking at the world in the last nine modal aspects that are uniquely human. Each design norm points to an important dimension of the technology design space; each represents a certain facet of the structure of reality. The norms can help us to consider the big picture as we make choices about technology. The following subsections briefly describe each design norm, point to a scriptural basis for the norm, and then suggest ways it can guide technology design.

The analytical norm. The analytical aspect directs us to use wisdom and discernment when we use the resources of God's good creation to build technology. Although the authors of *Responsible Technology* did not explicitly highlight this mode as a design norm, it underlies much of their discussion about the philosophy of design. For example, specifying a design, analyzing tradeoffs, and determining the effectiveness and sufficiency of the design all depend on deductive reasoning, conceptualization, and even imagination. In his writings, Egbert Schuurman often includes the norm of effectiveness, which he distinguishes from stewardship. To be effective, a designer needs to analyze both what is needed to arrive at a distinct definition and a metric for what we mean by *effective* and then determine what is needed to achieve that goal.

Dooyeweerd's analytic aspect does not simply involve mathematics and computation but also the ability to make distinctions. Human wisdom is required to categorize both iron and copper as metals and also to distinguish them based on their strength, ductility, conductivity, and other characteristics. Furthermore, analyzing requires creativity, allowing us to conceptualize new ideas—conceiving of an alloy such as steel that is stronger than the iron from which it is made, for example. The analytical norm lets us appreciate God's creation by recognizing its diversity and seeing patterns of commonality in categories of elements and in families or species of living things. Furthermore, when used best it encourages us to imagine how to cultivate and expand creation, enabling and supporting all the other norms.

The ability to conceptualize, distinguish, think logically, and plan ahead are all part of the analytical norm. These actions, however, require more than mere knowledge. We can creatively draw inferences and insights from data, but we also need wisdom to do so. Wisdom is more than analysis, but wise discernment comes through thoughtful analysis. Our modern society tends to focus on information and data, pretending that mere knowledge is wisdom. The poet T. S. Eliot wrote about this prior to World War II, well before the information age had fully developed:

> All our knowledge brings us nearer to our ignorance,
>> All our ignorance brings us nearer to death,
>>> But nearness to death no nearer to GOD.
>>> Where is the Life we have lost in living?
>> Where is the wisdom we have lost in knowledge?
> Where is the knowledge we have lost in information?[9]

The teacher in Proverbs exhorts us to seek wisdom, emulating the ant that sets aside provisions for future needs: "Go to the ant, you sluggard; consider its ways and be wise! It has no commander, no overseer or ruler, yet it stores its provisions in summer and gathers its food at harvest" (Proverbs 6:6-8). The teacher also exhorts us to avoid the lazy idleness that does not think ahead: "Sluggards do not plow in season; so at harvest time they look but find nothing" (Proverbs 20:4). For the psalmist, wisdom, at least in part, rests on careful thought and analysis. "Teach us to number our days, that we may gain a heart of wisdom" (Psalm 90:12). Wisdom leads to the underlying truth of the matter. "Buy the truth and do not sell it—wisdom, instruction and insight as well" (Proverbs 23:23).

The cultural norm. The cultural norm is based on the formative modal aspect. It is associated with shaping history, forming culture, and pursuing goals and achievement. Consequently, it speaks to technology, which is one kind of cultural activity.

I was fascinated by technology from an early age. When the TRS-80 computer was released by Tandy in the 1970s, I would hang out in the local RadioShack store just to get my hands on one. At first, I tried playing the simple games they had running as a demo. Thanks to a tolerant store manager, I soon discovered I could write my own programs and began teaching myself BASIC. The manager did not let me save my programs, however, and he would shoo me away when an adult customer wanted to look at the computer. When the machine was free again, I'd start over creating my program. While I had friends that could draw better, sing more beautifully, or sink a free throw more gracefully, this was a domain where I creatively excelled. My programs were crudely simple at first, but they were my artistic expression.

Over the years I learned many more programming languages, including FORTRAN, LISP, Perl, Python, Java, and sailed every programming sea: C, C++, C#, and Objective-C. Each of these programming languages conveys meaning, similar to human languages. We write programs to communicate our will to the computer, and we communicate with fellow programmers who may need to make adjustments or additions to the program. Thinking about it this way can help us see that writing a software program is a type of cultural formation. A software developer can create a program of beauty and elegance or one that is ugly and difficult to understand.

Dooyeweerd considered formative power to be the kernel of the formative modal aspect. Kalsbeek adds to our understanding of this aspect by describing it as "the controlled shaping of material according to a free design."[10] The shaped material could be something physical, which is why Dooyeweerd felt technology was strongly represented in this aspect. We humans shape the world around us. We are material shapers and tool makers. We are designers by design. God endowed us with his image, including creativity and the ability to cultivate the creation.

Figure 4.1. Radio Shack TRS-80 computer

Designing buildings is one way we form culture in response to God's call to unfold and cultivate his creation. This even includes respecting past construction. Architects who design new buildings to be consistent with old buildings nearby are honoring their cultural history while they are creating new culture. The cultural norm also includes respecting building codes, which represent what the community has learned about safe and dangerous practices. Buildings, like all technological products and processes, are cultural artifacts and a form of cultural formation. That is why Dooyeweerd sees the invention, design, and production of technology as an example of the formative modal aspect.

Designing buildings also can destroy, corrupt, or undo culture. This happens when we neglect human relationships in the social realm and neglect our task of cultivating the creation in the natural realm. Destroying a historically significant home in order to build a nondescript office building would be an ill-considered attempt at cultural formation; demolishing a decrepit old building that endangers the neighborhood to build a new community building would positively shape a community. Technology that destroys other cultural features is equally ill-considered. For example, social media can amplify voices of hate that encourage the breakdown of family relationships, the destruction of art, or the disintegration of the rule of law by legitimate governing authorities.

Designing technology is a creative, cultural activity that produces concrete cultural artifacts, drives historical development, and communicates deep cultural meaning. As a culture-forming activity, the engineer or scientist who produces technology is analogous to the composer who writes music, the writer who pens a novel, the senator who introduces legislation, or the activist who leads a social movement. The act of producing culture is a distinctly human activity with biblical roots. Let's take a closer look at its scriptural foundation next.

In chapter two, we discussed the significant implications of the very first words that God spoke to the humans he created. He commanded humans to take care of the creation and made them his stewards: "God blessed them and said to them, 'Be fruitful and increase in number; fill the earth and subdue it. Rule over the fish in the sea and the birds in the sky and over every living creature that moves on the ground'" (Genesis 1:28). This directive can be read broadly as a command to unfold the gifts that God has given us in the myriad materials that can be formed and shaped. To be clear, we are not called to simply keep the creation in its original form, though there is much merit in this. We are also called to change it and produce new things. Psalm 104:14 says that God makes "plants for people to cultivate." Cultivation requires tilling the soil, selectively caring for and nurturing the plants. But our mandate is not exclusively agricultural. Isaiah prophecies of a coming time when God's people will "build houses and dwell in them" (Isaiah 65:21). Stewardship is caring for the creation both as it is and as it should become.

Each of us is called by God to be a steward for our corner of the earth. That corner might be a geographical space, such as the garden that Adam was charged to care for: "The LORD God took the man and put him in the Garden of Eden to work it and take care of it" (Genesis 2:15). That corner might also be a figurative space, such as the discipline of engineering or the responsibility of parenthood. God is sovereign over all creation. Even after creation was darkened by sin, Christ's death and resurrection provided a redeeming light. As Christians working in technology, we are accountable as stewards fulfilling the cultural mandate in the gadgets we invent and accountable as redemptive agents in claiming Christ's reign over technology.

The norm of clarity. Dooyeweerd's lingual aspect focuses on communication, such as symbols, that convey meaning—bringing clarity. For designers, this means communicating in ways that bring transparency and shared understanding.

The norm of clarity demands that writing is clear and speech is meaningful, effective, and truthful. Honoring God through this norm drives us to seek mutual understanding by clear transmission of meaning. Dishonoring God comes from employing messaging that purposely distorts, lies, manipulates, or obscures true meaning.

For an example of a clear design, let's look at the disc brake system in an automobile. When a driver presses the brake pedal, pressing harder causes the car to brake harder. Warning lights indicate a loss of brake fluid. And the brakes use a special wear indicator tab designed to produce a screeching sound when the pad is getting too thin. Both safety and rotor integrity can be compromised if the pad is not replaced before it fails. All of these features help the user understand key aspects of the design that relate to use, safety, and maintenance.

WEAR INDICATOR TAB

Figure 4.2. Automobile disc brake

As citizens in a democratic society, we demand transparency from our government. Accurate and complete information prevents corruption, enables wise decisions, and provides accountability. The norm of clarity also applies to technology. Each stakeholder needs access to accurate information in order to prevent abuse of the technology, make wise decisions about its appropriate use, and offer accountability.

> **DEVICE PARADIGM** is a term from philosopher Albert Borgmann that describes technology as mechanisms that procure commodities while hiding their inner workings.

Technology does not always use symbols that convey clear meaning. Technology philosopher Albert Borgmann has described this problem as the "device paradigm."[11] He saw that many of our modern technological products were commoditizing some benefit but at the same time obscuring the means to achieve that end. Rather than technology as a tool or instrument, technology as a device hides the inner workings in a black box that mysteriously produces a commodity such as heat, motion, or information. The furnace is an example, providing heat at the effortless touch of the thermostat, replacing the focal point of a fireplace where family gathered to keep warm and eliminating the need to chop firewood. With these activities no longer needed, the family lost a rhythm that helped them bond and strengthen their relationships.

Clarity enabled by transparency can also support other norms. For example, open-source software encourages community development of applications because the source code is not restricted by proprietary licensing. This gives developers broader freedom to develop, thus also honoring the cultural norm.

There may be occasional justification for confidentiality and secrecy with regard to proprietary trade secrets and intellectual property. One reason might be to protect the rights of the creator, which also promotes the norm of justice that is discussed later. However, confidentiality could become an excuse for promoting deceit. Protecting intellectual property might be used as a shield for those that create and sell products, enabling them to hide certain aspects of the design even if they are necessary details for the user to understand the value, risks, safety measures, and hazards related to the technology. The challenge is thus to balance the goal of transparency for the user with the goal of protecting the rights of the developer.

It is important to note here that more information does not always give more clarity. We've all seen safety notices that go on for dozens of pages. They are difficult to digest and almost always get tossed out with the product packaging without a second look. Information overload detracts from, rather than enhances, clarity.

Earnest pursuit of the norm of clarity promotes truthfulness and eschews deceit. Scripture addresses these themes frequently: "Do not lie to each other, since you have taken off your old self with its practices and have put on the new self" (Colossians 3:9-10); "The LORD detests lying lips, but he delights in people who are trustworthy" (Proverbs 12:22); "An honest witness does not deceive, but a false witness pours out lies" (Proverbs 14:5); "Do not spread false reports" (Exodus 23:1).

The social norm. Dooyeweerd's social aspect builds on the lingual aspect to address interactions between humans. Teamwork, polite courtesy, fellowship, friendship, and association are all examples of social norms.

Teamwork is emphasized in engineering for good reason. It ensures that any design problem is approached from multiple perspectives. A diversity of viewpoints increases the number of potential solutions, provides a double-check that all the angles have been considered, and helps balance competing needs and goals.

Social courtesy is the first step toward peaceful relationships—though a deep-seated peace also requires justice, a norm we shall discuss further below. Courtesy toward one another is one way we show respect for all persons; it also demonstrates a willingness to consider and address everyone's ideas, goals, needs, and fears. Technology that encourages positive human relationships honors this norm. Technologies that create isolation, hatred, and segregation dishonor this norm. Many technologies do both, of course, and part of our job as redemptive agents of Christ is to work toward enhancing the one and suppressing or eliminating the other.

Let's apply this idea to social networking technology. Examples at the time of writing this book include Facebook and Snapchat, but you may be using something entirely new by the time you read this. These technologies make it easy to quickly post an idea or image, which one can quickly regret if done impulsively. Unfair and unfounded criticism that is not carefully thought out can lead to hate and anger. However, these technologies can also help families and friends that are separated geographically to connect, showing love and encouragement to one another.

At the time of this writing, the Covid-19 pandemic prevented the authors from meeting together in person periodically as we originally planned. Video web conferencing technology provided a way to still talk with one another, including visual connections. Seeing body language is important so that you can more easily tell if your words were taken well or poorly by the hearer. While Covid-19 forced many workers to stay home, a good portion could continue working using a variety of electronic communication tools. This would not have been possible for most people even a short decade ago, but the technology has come along far enough to support productive work, a blessing that reduced the financial impact of Covid-19 to some extent. As an elder in my church, I could not meet with the members in my elder care group during the pandemic, but we were able to come together virtually on a video call to encourage one another, read Scripture, and pray together.

The norm of stewardship. The design norm of stewardship is based on the economic aspect. Although the term *economics* as used today generally brings to mind financial laws of supply and demand, the historical meaning of *economy* is the careful management of limited resources under one's care. We prudently manage our resources when we use them wisely and frugally, not frivolously. Christians understand that the resources that we oversee are not our own. We are caretakers and stewards who oversee our master's property.

I come from Dutch descent, an ethnicity sometimes stereotyped as frugal to the point of penny pinching—though we are not the only ones! Insofar as pinching pennies implies a greedy or miserly attitude, it is not the type of frugality promoted by the norm of stewardship. Rather, the idea is that we have a responsibility to use wisely the resources that God has entrusted to our care. We must not hoard them but use them as a way to advance God's kingdom. Specifically, we should measure the return on our investments of resources not by how we profit but by how much we have glorified God and enhanced the welfare of our neighbor.

Cost is always a design criterion for technological design activity. Anyone who has ever claimed that cost was no object has never employed a creative engineer who can always imagine a lavishly expensive design solution. Because designs that reduce costs often make the most careful use of resources, efforts to minimize cost are often prudent. However, we must be on our guard not to abuse a resource that is not counted in the cost. For example, designs that pollute air, water, or other aspects of our environment may not fully incorporate the cost of their impact unless external regulation forces us to account for such impact.

The norm of stewardship may sometimes persuade us to make design choices that actually increase the cost of the design, at least at first glance. And if we find that this makes our design less competitive in the marketplace, we may be called to speak prophetically. We should preach that appropriate design standards and regulations must properly count all costs to society in order to create a level playing field so that designs thought to be more expensive financially in the present are now perceived as frugal in a broader sense. Being stewardly and prudent does not mean simply using the fewest possible resources. Instead, we should think of stewardship as the cultivation of a garden. When irrigating a garden, we do not waste water and flood the plants, but neither do we skimp on the water so much that our plants become parched. Madison (our intrepid web developer) had been liberal in her creative visual web designs, but she had not cultivated her web garden in a way that could delight her visually challenged customers.

The idea of God charging humans to be stewards dates back to the first chapter of the Bible: "God blessed them and said to them, 'Be fruitful and increase in number; fill the earth and subdue it. Rule over the fish of the sea and the birds in the sky and over every living creature that moves on the ground'" (Genesis 1:28). We read that after God finished the creation, he delegated authority to care for that creation to humans. Our authority as stewards, however, does not give us license to plunder resources or indulge ourselves. Our authority is a sacred responsibility. God puts us to work. "The LORD God took the man and put him in the Garden of Eden to work it and take care of it" (Genesis 2:15). The story of the ant as an example of wisdom in Proverbs 6:6-8 demonstrates the careful planning that stewardship of resources requires, demonstrating that stewardship builds on the earlier norm of analysis.

The norm of harmony. The harmonious design norm builds on the aesthetic aspect. Technology, like everything else in life, can exhibit beauty, harmony, an integrity of form and function, even surprising delight and playful enjoyment. In music, for example, harmony playfully complements the tones and timing of the melody. Although musical harmonies are meant to be heard, the relationships can be spotted visually within the sheet music even by amateur musicians. Harmonious beauty in a device is not merely cosmetic, it must marry form and function so that it is intuitively a pleasure to use. A harmonious hammer has the right balance and grip that allows you to pound nails effectively. A harmonious smartphone fits your hand to make tapping effortless and provides a visual interface that is intuitive.

Beauty is praised in the Bible, but the context is always essential for understanding what is praiseworthy. The tabernacle and, later, the temple are both described with details that help us imagine their beauty, but they also function as the house of God. Paul tells the Philippians to think about whatever is lovely, but it is part of a list of praiseworthy attributes that includes whatever is true and right (Philippians 4:8). Beauty by itself, if it is not accompanied by other admirable qualities, is not admirable but merely shallow. Form without function loses integrity and becomes ugly. For example, Jesus criticizes the Pharisees because they, like whitewashed tombs, are "beautiful on the outside but on the inside are full of the bones of the dead and everything unclean" (Matthew 23:27). Isaiah describes feet that are beautiful because they are bringing good news (Isaiah 52:7). The design norm of harmony puts these lessons into practice for the designing of technology. Designs that are functional but ugly might be acceptable, but designs that are functional and simultaneously lovely are admirable. Designs that are cosmetically appealing yet dysfunctional should be avoided.

Achieving both function and form is not easy, requiring technical know-how and artistic flair. An example method for pursuing both is a User Interface (UI) design technique from Apple, a company well known for beautifully fitting together form and function. User interface designers thinking about a new idea for an app take a sheet of paper that has the familiar rounded rectangle outline of a smartphone, repeated ten times in a grid on the sheet. They now make a pencil sketch in the first outline to represent how the main interface might appear. Next, they must come up with a different way to show it, drawing it in the second phone outline, and so on. The first few variations are relatively easy, but coming up with ten can be challenging. Yet, the last variations are often the ones that bring form and function together in a pleasing, intuitive design.

The norm of justice. We all have an innate sense of satisfaction when justice is served and feel galled when justice is frustrated. Justice is served when everyone gets what they deserve. Justice is served when everyone receives what they are due, when rights are respected, when responsibilities are carried out. One of Kafka's stories starts out with the line: "Someone must have been telling lies about Josef K., he knew he had done nothing wrong but, one morning, he was arrested."[12] We immediately recognize the injustice of wrongful imprisonment.

We know in our hearts that injustice is wrong. If you have experienced injustice, you know what it is like to cry to the Lord in anguish seeking relief. If you love someone who has experienced injustice, you know the heartache of seeing them wronged. A community that knows injustice cannot know peace. It's not only our hearts that tell us justice is right; we also learn this from Scripture. God clearly demands that his followers pursue justice. "Blessed are those who act justly, who always do what is right" (Psalm 106:3). "Evildoers do not understand what is right, but those who seek the LORD understand it fully" (Proverbs 28:5). Engineers understand requirements. Micah tells us what God requires of us: "He has shown you, O mortal, what is good. And what does the LORD require of you? To act justly and to love mercy and to walk humbly with your God" (Micah 6:8).

Some of the demands of justice are obvious. Professional ethics and legal regulations already hold engineers responsible for the safety of the users of their designs. But Christian engineers are called even more to seek justice for our users through our designs. This is a daunting responsibility. While our personal actions might create injustice for one or two others, our professional acts in designing technology have the potential to create injustice for large groups of people. Predicting the full social and cultural consequences of new technology is by no means easy. The job is made even more difficult because of internal biases surrounding technological development today. Because costly technology development is usually funded by the wealthy, we need to carefully balance the rights of stakeholders (i.e., those impacted by the technology) who cannot fund a technology with the needs and desires of the paying customer.

An example may make the need to consider justice in technology more clear. Consider the bias of automobile design that puts certain passengers at greater risk: women are 17 percent more likely to be killed in a car crash than men.[13] This difference is embedded in the design of cars and can be traced to the fact that crash testing rarely represents both genders. "Even though female and male bodies react differently in crashes, an average adult female crash test dummy simply does not exist, despite the fact that women obviously drive to work, take road trips, and ride in cars with friends. That absence has set the course for four decades' worth of car safety design, with deadly consequences."[14]

Technology contributes to justice or injustice, intentionally or unintentionally. All engineers should strive to anticipate the consequences of their design decisions since unanticipated consequences can bring harm or even fatalities. Intentionally designing technology to create an injustice is malicious, likely driven by greed or desire for power. Consider the bridge designs of Robert Moses (discussed in chapter three) that were intentionally designed to segregate.

Even if the injustice is not recognized or anticipated at first, once it becomes apparent, it becomes intentional and must be addressed if we are to follow God's call to seek justice. Failure to do so with reasonable care is negligence. Being consciously discerning about the future consequences of a design helps the designer be on the lookout for consequences that might unfairly impact one group of users disproportionately. For example, software programmers trying to determine credit scores for a bank might think using easily accessible data such as zipcodes could act as a proxy for creditworthiness, but such a design choice could easily result in perpetuating cycles of poverty.[15]

You might be convinced that it is our professional duty to anticipate how users might interact with our designs. You might also be convinced that Christian faith demands justice. However, you might not see that the two are linked. Remember the story from the previous chapter about God requiring a parapet around the roof. If God expected the builder of a home to anticipate safety issues for guests, then it seems that God would also expect technology designers to anticipate safety issues for our guests—that is, the customers, users, and stakeholders of our technological products. It is admirable to avoid injustice in a design. Actively seeking justice through designs is even better, though, because we know that our God loves justice.

Seeking justice through technology means that we not only attempt to anticipate negative consequences but that we try to enhance the cause of justice. Technology that helps prevent crime or helps law enforcement identify wrongdoers, for example, enhances justice. We must beware of oversimplifying, however. Technology in the hands of the state does not always yield justice. For example, powerful surveillance technology can become overly intrusive and can be easily abused, creating new injustice.

Earlier in this book, we've come to know the amplifying effects of technology. Designers can steer that amplifying power toward the needs of those who are least fortunate, giving them a better chance at health and success. This is important. Although justice demands fair treatment for all people, God also calls us to keep special watch for the oppressed, who may not get their rights unless someone seeks justice on their behalf: "Stop doing wrong. Learn to do right; seek justice. Defend the oppressed. Take up the cause of the fatherless; plead the case of the widow" (Isaiah 1:16-17).

Justice-enhancing technology on an individual scale might include the design of advanced hearing aids to amplify foreground conversational voices for those with hearing loss while using digital signal processing to filter out background noise.

Justice-enhancing technology on a city-wide scale might include a roadway system designed to safely accommodate cars, buses, bikes, and pedestrians. A public transportation system might be expanded beyond the bus to include light rail to reduce traffic congestion. Crosswalk signals might be enhanced to include not only a visual signal to indicate safe passage but also audio signals for the visually challenged.

Justice-enhancing technology on a state-wide scale might be an unemployment benefits enrollment system that does not require the unemployed person to have a computer in order to apply, providing alternative methods including a telephone-based system.

The norm of caring. The design norm of caring is based on Dooyeweerd's ethical aspect. It draws on the virtues of generosity and charity, and ultimately self-giving, sacrificial love. The parable of the Good Samaritan gives a biblical paradigm for this norm: "He went to him and bandaged his wounds, pouring on oil and wine. Then he put the man on his own donkey, took him to an inn, and took care of him" (Luke 10:34). Even in this ancient story, we find a model of how technology can be directed toward caring. The oil, wine, and the inn are all early technologies, which today might be considered chemical and civil engineering products. They are the tools that the Samaritan put to work in the service of caring. They are tools for helping us love our neighbor.

As we've noted many times, caring for those around us is a central calling of Christians, a commandment that Jesus tells us is a summary of the whole law: "And the second is like it: 'Love your neighbor as yourself'" (Matthew 22:39). As Christians working in technology, we have a unique opportunity to develop technology that helps care for our neighbors, even globally. For example, the engineer who takes a vacation from her day job to work in a mission field and help install a new well for a village is showing the love of Christ to that village. Spending time designing a more effective well and filtering system that could benefit thousands of villages brings such help even further—or designing the system and then participating in fielding it, in order to see the direct benefit to clients.

Madison, the web developer who had prioritized a striking visual design, might have realized the need to consider visually impaired users if she had also prioritized the norm of caring. Like modern-day Good Samaritans, engineers can use their technological skills to care for the less fortunate. However, one challenge to technological caring is that technology is often too expensive for the neighbor who needs it the most. Our consumer-oriented society tends to target what will produce the most profit. This means that we often tune our technology toward the rich.

Caring for the poor requires that we also pursue technological innovations that are affordable. These are not mutually exclusive pursuits. Consider the mobile phone that at first could only be afforded by the wealthy. Over time, engineers reduced costs through clever design improvements and by producing them in higher volumes. Eventually, cellphones began appearing throughout society and in the majority of the world's nations, actually opening up communication opportunities for economically poor people. Another example relates to medical equipment and designer drugs that are often initially so expensive that only the wealthy can afford them. Even seemingly commonplace medical devices such as eyeglasses can be out of reach of those living in extreme poverty. However, talented engineers are addressing this need by continuing to develop techniques and equipment, such as self-adjusting eyewear that can provide visual assistance for people in remote locations where an optometrist is often not available.

The norm of faithfulness. The final design norm builds on the modal aspect that relates to faith, though Dooyeweerd used the more philosophical term *pistic* as the name of this aspect. This norm describes our worldview—our basic commitments, presuppositions, and beliefs.

As Christians, our ultimate loyalty must be to God and our ultimate trust must be in God. Engineers must be on guard that technology does not lead users to put too much faith in technology to solve all of humanity's problems (more on this in chapter six). "Some trust in chariots and some in horses, but we trust in the name of the LORD our God" (Psalm 20:7). There is a paradox here: even though we aim for high reliability in our designs so that our users can trust they are safe, encouraging too much trust and reliance on technology allows people to make it an idol. The more technology gives users control over their environment, the more it can fool them into believing they are the masters of their own destiny—that they are in control, rather than recognizing that God is in control. Successful designs that gain the trust of the users of our technology can also lead designers into the sin of pride. Simply put, unless designers consciously choose to follow the norm of faithfulness, our designs can lead us and our users into unfaithfulness.

One way we observe the norm of faithfulness is by recognizing that no technological device can be perfect—and then moving ahead with humility. The infamous failure of the Tacoma Narrows Bridge in 1940 was perhaps an example of hubris by engineers who designed bridges to span larger and larger gaps with thinner and thinner decks without fully understanding the impact of wind dynamics. When we build in failsafes, warning lights, and redundancies, we recognize that we cannot completely trust our creations. Humility gives us a sober view of ourselves and our technology, and it helps us and our users turn our eyes away from ourselves and our devices and instead turn them to God.

Figure 4.3. Tacoma Narrows Bridge collapse, November 1940

Faithful design can also deepen our faith and worship. For example, calendar apps can remind us to read God's Word daily; microphones amplify the pastor's voice; T-coil hearing loops in the sanctuary allow even those with hearing loss to participate (also honoring the norm of caring). As I was finishing this chapter during the coronavirus lockdown of 2020, many churches canceled in-person worship services and used live streaming and video conferencing to stay connected and worship together as a community even while separated physically.

CONCLUSION

This has been a long chapter, so a quick summary list of the design norms might be helpful to sum it up:

- ▶ Analytical—use wise discernment in design
- ▶ Cultural—make technology choices appropriate to culture and society
- ▶ Clarity—openly communicate the benefits and dangers of technology
- ▶ Social—honor and enhance relationships
- ▶ Stewardship—make frugal (but not stingy) use of resources
- ▶ Harmony—design technology aesthetically with integrity of form and function
- ▶ Justice—treat all people fairly with technology design, avoiding discrimination
- ▶ Caring—show love and kindness by helping others through technology
- ▶ Faithfulness—honor God with technology and put ultimate trust in God

Is this list of norms complete? Perhaps, but even Dooyeweerd was not certain his list of modal aspects was complete. There is no one answer or list for how to live our lives and no one answer for how to design our technology. The primary utility of these lists might be to provoke our thinking in directions we might not naturally take. Discernment is always necessary. We need to be careful not to simply choose answers or individual norms that suit us but to seek God's guidance as we work.

You may be wondering at what point in the design process one should apply these norms. We can summarize most technology design cycles as the four steps of (1) problem definition and requirements specification, (2) alternative solution ideation, (3) analysis to converge to a selected solution, ending in (4) implementation/validation. The engineering philosopher Peter-Paul Verbeek and others have suggested additional steps between stages two and three, each with normative implications: (a) explore design options and ideas for technology, (b) imagine and ideate to discover the normative implications of ideas, and (c) reflect and reframe to consider what kind of society would come from these ideas. We may need to step back even further and consider that the problems we choose to solve should be selected with the norms in mind.

In developing new technological designs, engineers will always find themselves choosing between design alternatives. Design choices, like most decisions in our lives, involve tradeoffs. Improving one criterion or goal often diminishes another. We can make a car go faster, but the fuel efficiency drops; we can make a car safer with heavy materials, but the speed and fuel efficiency drop. Because we are fallen and finite creatures, tradeoffs might even be necessary between the norms we've listed. We might find that we can improve justice for a disadvantaged group, but to the detriment of stewardship. We might find that we can improve caring for people that are ill, but at the cost of radically changing their cultural identity. This is not the way it is supposed to be. This is not God's shalom. But in a world broken by sin, it is not always possible to find a completely faithful response. Still, we are called to honor God, pursue righteousness, and strive for the best designs we can. We must not despair or give up. We must rely on God for creativity and discernment as we pursue what we believe will best help us love God and love our neighbor.

CHAPTER FIVE

BEYOND ENGINEERING ETHICS

Your hands made me and formed me; give me
understanding to learn your commands.

PSALM 119:73

No field guide to technology would be complete without addressing the topic of ethics. In this chapter, Derek Schuurman sketches a brief introduction of common ethical frameworks, but also seeks to broaden our view of the ethics landscape beyond hypothetical case studies to implications for all of life. This chapter returns to the biblical story and builds on the more comprehensive set of norms described in the previous chapter.

When Julie Wilson began working for a company that designed and manufactured medical devices, her employer asked her to review a respirator that would be used with infants. As she looked into the design, she discovered a flaw in the placement of a pressure relief valve meant to prevent excessive air pressure from being applied to an infant's lungs. Under certain conditions, the machine design could expose an infant to a dangerous level of air pressure. Julie recommended a simple repositioning of the pressure relief valve. Later, she discovered that the design change had not been made. Hundreds of respirators with the faulty design had been shipped and were being used in hospitals. Julie faced a difficult decision about what to do—one that could lead to a clash with her employer, but more importantly, one that had real consequences for the safety, health, and welfare of the public.[1]

This story is one of numerous examples that can be found in ethics textbooks and journals. Engineers, because of the nature of their work, may encounter situations related to public safety. They have an ethical responsibility to exercise a high standard of care with their technical expertise, particularly when safety is involved. Engineering ethics is generally part of the engineering curriculum and is typically taught by introducing some classic ethical frameworks. Later, students apply what they have learned by examining a variety of case studies, like the one given above.

Codes of ethics have been published by several engineering societies and organizations to help guide their members. Some examples include the National Society of Professional Engineers (NSPE), Institute of Electrical and Electronics Engineers (IEEE), Association for Computing Machinery (ACM), and American Society of Mechanical Engineers (ASME), each of which has a code of ethics for members. The purpose of this chapter is not to replicate materials that are already available, but to think about engineering ethics in the wider context of a biblical worldview, taking into account the norms discussed in the previous chapter.

Ethics is a branch of philosophy that explores questions of human morality such as right and wrong conduct. In engineering ethics, questions of morality are often explored in the context of thorny case studies in which an engineer must apply some ethical principle to determine a course of action. The problem with using case studies alone is that it can lead one to think that right and wrong conduct only arises in special ethical dilemmas—with the implication that the rest of engineering is neutral. The truth is, in the words of C. S. Lewis, "There is no neutral ground in the universe."[2] Everything we do, each design decision as well as our attitudes and our treatment of colleagues and customers, flows from our heart and is informed by religious commitments. In this sense, *all of life is religious.*

This chapter begins with a brief review of some ethical frameworks commonly used in engineering and shows how they relate to biblical principles. We conclude the chapter by situating engineering ethics within the broader normative framework introduced in the previous chapter and within the larger biblical story.

ETHICAL FRAMEWORKS: A BRIEF REVIEW

In general, common frameworks for ethical decision-making include *deontological ethics, consequentialism,* and *virtue ethics.* Each of these frameworks prompts different questions and highlights different concerns as summarized in the following sections.

Deontological ethics. One ethical framework is *deontological ethics,* an approach that focuses on following rules, laws, and obligations to guide our actions. Following rules and principles is appealing to many engineers who are accustomed to doing this in their technical work. But rules and principles only take you so far with ethics since there cannot be a rule for each unique circumstance and context.

Most engineering codes of ethics include elements of deontological ethics. For example, the NSPE code of ethics includes "rules of practice" and "professional obligations" sections that begin with the phrase "engineers shall . . ." followed by a variety of rules and requirements.[3] These rules set boundaries and guidelines for professional conduct. It is not uncommon for engineers to encounter situations where one of these ethical rules or regulations could be directly applied. For instance, the NSPE code includes a rule to "not disclose, without consent, confidential information," a straightforward rule that I have had to observe in my work in industry.[4] However, not all real-world situations can be addressed by specific rules because each situation has its own context and nuance. It requires wisdom and "sanctified common sense" to apply rules wisely to situations that arise but may not be specifically addressed.[5]

The Bible includes rules and laws—the Ten Commandments, for example. In Matthew 22:37-40, Jesus summarizes the Law and Prophets in the two commandments to love God and to love our neighbor. These two commandments point to a key for getting beyond ethical rules: our ethical thinking must be informed by love. In fact, without love, it is still possible to follow rules meticulously, but with corrupt motives and intent.

Consequentialism. A second ethical framework is *consequentialism*, the notion that the right thing to do is determined by whatever leads to the best overall outcome. Of course, engineers need to be concerned about overall consequences. For example, designing something to be more sustainable justifies increased cost and effort. However, consequentialism can also lead to the ends justifying the means. As Christians we cannot simply be concerned with outcomes; we also need to be concerned with our actions.

Utilitarianism, one of the most well-known forms of consequentialism, judges something to be good based on its utility. As long as the result is deemed "good," any means to achieve that end is justified. Sometimes the ends really do justify the means, such as when a student decides to apply hard work and effort in order to achieve a good grade.

Adopting utilitarianism as a standard for action, however, could lead the above student to justify *any* means, even cheating, as long as it resulted in a good grade. This approach considers any decision that results in greater good for the greatest number of people to be the proper choice. A classic example of bluntly applying this rule is that of the doctor who has five sick patients, each with a different ailing organ. By harvesting five organs from a perfectly healthy patient, the doctor saves five lives at the cost of one. But this cannot be right. Another example arises in certain autonomous vehicle scenarios when avoiding a road hazard to save the lives of the occupants of a car might result in killing a pedestrian.[6] Ethical choices can be perplexing, but they should not be reduced to quantifying utility to determine whether an action is right or wrong.

Utilitarian approaches are also challenging because we cannot always predict with confidence the outcome of an action. Furthermore, we may not all agree about what is "good" for the greatest number. Even if we could agree on what is "good," there are often unintended consequences that we will fail to anticipate.

The NSPE code of ethics also includes a concern for consequences. Engineers are called to "hold paramount the safety, health, and welfare of the public" since engineering decisions can have tremendous consequences. Furthermore, the code encourages sustainable development with a goal "to protect the environment for future generations." However, while the goals of safety and sustainability are important, the means by which these are pursued must be shaped by other requirements laid out in the code of ethics.

Christians certainly are called to consider consequences and to live in a way that promotes shalom (as discussed in chapter two). However, we are not called to impose shalom by any means that we see fit, rather we are called "to act justly and to love mercy and to walk humbly with your God" (Micah 6:8).

Virtue ethics. A third approach to ethics is known as *virtue ethics* and focuses on moral character rather than on rules or the consequences of actions. Virtue ethics emphasizes becoming a virtuous person—developing attitudes, dispositions, and character traits that enable one to respond wisely in a particular situation. Ancient Greek philosophers identified four cardinal virtues: prudence, justice, temperance (self-control), and courage.

These four virtues each appear in some form in the NSPE's code of ethics. The NSPE alludes to prudence by calling engineers to exercise care and good judgment. Prudence is a helpful virtue for keeping the NSPE requirement to "hold paramount the safety, health, and welfare of the public." The NSPE code requires engineers to practice justice by stating that "the services provided by engineers require honesty, impartiality, fairness, and equity." The code affirms the virtue of temperance (which deals with moderation and self-control) by requiring engineers to "conduct themselves honorably, responsibly, ethically, and lawfully." Finally, the NSPE code calls engineers to do things that require courage, such as "acknowledge their errors" and "advise their clients or employers when they believe a project will not be successful."[7]

The Christian tradition includes other virtues that were not highly valued by the Greeks.[8] In 1 Corinthians 13:13 Paul writes, "And now these three remain: faith, hope and love. But the greatest of these is love." Faith, hope, and love (sometimes referred to as the *theological virtues*) are important themes in Christian thought and action. Faith deals with where we place our trust. Instead of placing our trust in technology, Christians need to fix our eyes firmly on Christ. Second, we need to maintain hope. But our hope must not be placed in a secure job, a lucrative income, or the power of technology. Instead, it must rest on God's faithfulness with an unwavering confidence in his promise to restore all things. And love reminds us of Christ's example and his call to love God and to love our neighbor as ourselves (Matthew 22:37-40). As engineers, we can live out the call to love God and neighbor by putting technology, from microelectronic circuits to megaprojects, in the service and practice of love.

Humility, another Christian virtue, is also helpful in engineering. Humility involves modesty—not thinking too highly of oneself and instead thinking more of others. Humility allows you to admit when you are wrong. Scripture speaks often of the importance of humility (see for example, Psalm 25:9, Proverbs 3:34, and Ephesians 4:2). As creatures our abilities have natural limits. Our strength, our memory, our knowledge, our intellectual aptitude, and every other capability we possess is limited. As we work with technology, we cannot possibly anticipate every possible scenario for our designs. Humility enables us not only to modestly and frankly assess ourselves and our human finitude, it also enables us to acknowledge the limitations of our designs. For example, an engine that is properly designed and cared for should normally not overheat. However, because we cannot anticipate all problems, we include a temperature sensor and an engine warning light so that drivers will know when something is not working. Likewise, when we use models in our engineering designs, humility helps us appreciate the limits and assumptions of our models and simulations.[9] Humility helps us welcome and regularly employ peer reviews to improve designs so that others may recognize problems that might have become blind spots for us. In these ways, humility can make us better engineers.

CHRISTIAN CHARACTER

The word for *virtue* does not appear often in the New Testament, but there is an emphasis on "the careful development and cultivation of Christian character."[10] The focus on virtues actually originated with Aristotle and the ancient Greeks, but author and theologian N. T. Wright suggests that the Christian vision is "larger and richer" than theirs. The Greeks portrayed virtue as an individual moral effort, but the Christian is motivated by citizenship in God's coming kingdom. In the words of Wright, "Christian virtue isn't about *you*—your happiness, your fulfillment, your self-realization. It's about God and God's kingdom."[11] The virtues of faith, hope, and love all "point away from ourselves and outward: faith, toward God and his action in Jesus Christ; hope, toward God's future; love, toward both God and our neighbor."[12] In contrast to Aristotle's vision of the "moral giant striding through the world doing great deeds and gaining applause," the Christian vision is about seeking first God's kingdom.[13]

Christian virtues and character are developed through practice and only gradually become second nature. Consider the following story that Wright recounts in his book *After You Believe*. Captain Chesley "Sully" Sullenberger was flying an Airbus A320 out of LaGuardia Airport. About two minutes after takeoff, he unexpectedly ran into a flock of Canada geese and both engines were damaged and lost power. In the moments that followed, Captain Sully and his copilot had to rapidly make a decision about what to do as the plane began to lose altitude above a populated area. The options were limited. If they attempted to glide to the nearest airport, they could fall short and crash into a built-up area. If they attempted to land on the New Jersey

Turnpike, they could endanger cars as well as the occupants of the plane. One option remained: a water landing on the Hudson River, a difficult maneuver that could flip or break up the plane if not executed correctly. In the moments that followed, Captain Sully adjusted the plane's controls and speed to maximize their glide path and quickly determined that a water landing was the best option. Without power from the engines, he successfully turned the plane and expertly guided the plane to a safe landing in the Hudson River. All the passengers and crew escaped, something Captain Sully confirmed by walking up and down the aisle of the plane before leaving it himself.[14]

This event has been called "the miracle on the Hudson," but Wright makes the point that years of training and repeated practice gave Captain Sully both the skills he needed to fly a plane in a crisis and the virtues of "courage, restraint, cool judgement, and determination to do the right thing for others."[15]

What kind of habits, rituals, and practices help us develop Christian character? They include some ancient spiritual practices such as reading Scripture, regular prayer, Sabbath keeping, solitude, silence, and serving others.[16] Such practices help us to be like the "tree planted by streams of water" described in Psalm 1, a tree "which yields its fruit in season." In her book *Journey Inward, Journey Outward*, Elizabeth O'Connor makes the case that the journey out into the world must be sustained by the journey in—that is, an introspective journey.[17] Such practices spill out into the world, shaping our character and our work as engineers. And we do not do this alone; the sanctifying work of the Holy Spirit is also present in our lives, and the fruit of the Spirit yields "love, joy, peace, forbearance, kindness, goodness, faithfulness, gentleness and self-control" (Galatians 5:22-23). These ninefold fruits are not for our own benefit, but will impact those around us, also in the workplace. Indeed, character and conduct in the workplace will communicate far more than what we may say. Cultivating spiritual disciplines and practices and staying "in step with the Spirit" (Galatians 5:25) must be integral to the life of Christians, and hence the Christian engineer.[18]

BEYOND PROFESSIONAL ETHICS

As we said earlier, ethics is often taught by applying ethical theories to case studies to pick an action that would be best in a particular situation. The problem with this approach is that it is not necessarily clear what criteria one ought to use when choosing a particular ethical theory to apply. In hard ethical dilemmas, ethical theories may even come into conflict. Even so, most people sense the reality of right and wrong ways to act.

In his book *The Abolition of Man*, C. S. Lewis makes the point that human intuition can discern a "natural law," what the Chinese called the *Tao*, which transcends faith traditions. This *Tao* includes concepts that have been historically recognized in many civilizations, such as justice, beneficence, and magnanimity.[19] The fact that secular professional organizations recognize some kind of *Tao*, and hence a need for professional ethics, should not come as a surprise. Secular ethical frameworks reflect what Christians recognize as the reality of moral laws and norms established by God for his creatures. It is this foundational creational reality that provides a basis for Christians to engage our secular colleagues on moral questions at work and in the public sphere.

Wrestling with perplexing case studies can be academically stimulating, but they touch on only a small fraction of our lives. As Christians, we live *coram Deo*, before the face of God, and so we should always be asking, "How then shall we live?" The moral life involves seeing one's life as part of the biblical story and following its main character, Jesus Christ.[20] Philosopher Alasdair McIntyre puts it this way: "I can only answer the question 'What am I to do?' if I can answer the prior question 'Of what story or stories do I find myself a part?'"[21]

N. T. Wright provides a helpful illustration of how the biblical story can inform our lives in the modern world. He imagines that there is a Shakespeare play whose fifth and last act was lost. Because the first four acts are so engaging, a theater group decides that the play ought to be staged. They believe it would be inappropriate to write a fifth act and instead decide to give the play to "highly trained, sensitive and experienced Shakespearian actors, who would immerse themselves in the first four acts . . . *and who would then be told to work out the fifth act for themselves.*"[22] The actors would improvise the fifth act, informed by the story and trajectory of the first four acts. Wright suggests that this is how the Bible can be authoritative in our lives today. We are given the biblical story from creation to Jesus and the early church, and we know how the story is ultimately supposed to end. Like the Shakespearian actors, we are called to "improvise" by living sensitively into the biblical story in a way consistent with its trajectory within our current context.

In chapter two, we recounted the biblical story and the basic acts of creation, fall, redemption, and restoration. In the creation story, we read about how God's Word calls a diverse and complex world into being. We saw how the fall and its resulting sin and rebellion infected God's good creation, and how the redeeming work of Jesus Christ inaugurated his kingdom on earth, a kingdom that will be fully realized in the new heavens and earth at the end of time. In the meantime, God's people are called to act as agents of renewal, also in the vocation of engineering. It is within that grand narrative that we live and work. We are called to do this, not only as a matter of obedience but also as a matter of witness.[23]

Even though we are living within the story of Scripture, we can still make use and benefit from codes like the NSPE code of ethics. But these ethical codes, when seen within the biblical narrative, have a limited scope. For the Christian engineer, professional ethics are necessary but not sufficient. Our call as Christian engineers is far more comprehensive!

Most discussions about ethics focus on limited aspects of the norms covered in the previous chapter. For instance, the NSPE code of ethics points to the norm of justice, stating that engineers "shall not be influenced in their professional duties by conflicting interests." But the justice norm is much more comprehensive than avoiding injustice; it requires promoting and restoring justice in the widest sense. The NSPE's code of ethics alludes to the norm of clarity, stating that engineers should "avoid deceptive acts." But the norm of clarity goes much further than avoiding deception; it requires openness and honesty through clear and meaningful language as well as designs that intuitively communicate their intended purpose and associated risks. The NSPE's code of ethics touches on the norm of stewardship by calling engineers to "adhere to the principles of sustainable development in order to protect the environment for future generations." But the norm of stewardship has implications for not only *developing* sustainably but also *preserving*. Furthermore, it extends beyond environmental resources to include human resources. The norms outlined in chapter four not only go *deeper* than those in engineering professional codes of ethics, they are also *broader*. They include concerns for cultural appropriateness, beauty and aesthetics, love and caring, and faith and trust. All of these norms must be pursued simultaneously, not only to avoid harm but to be obedient to our call to act as agents of renewal in the larger context of God's coming kingdom.[24] Being attentive to all of these norms offers a wider vision of what is needed to help people and the rest of creation flourish.[25]

In essence, ethics can be summarized as discerning how to become the kind of person who does the will of God in a particular context. This requires character and virtue (the kind of people we are), discerning laws and norms (the will of God for his creation), and wisdom to understand the context of a situation.[26] In his book *Every Good Endeavor,* author and pastor Tim Keller connects how God shapes character in our personal and professional lives: "Here then is how the Spirit makes us wise. . . . He makes Jesus Christ a living, bright reality, transforming our character, giving us new inner poise, clarity, humility, boldness, contentment and courage. All this leads to increasing wisdom as the years go by, and to better and better professional and personal decisions."[27]

Focusing on case studies can serve to "bracket the realities of daily work" in order to focus on "situations that are comforting because they are so remote."[28] But our call as Christians is much more comprehensive than addressing a rare crisis. Consider the medical device case study at the start of this chapter. The details provided in this case study are few, focusing only on this particular incident. Clearly, Julie ought to urgently follow up on the placement and reliability of the pressure relief valve. However, normative considerations will also encompass the context of her daily work designing medical devices.

For example, the cultural norm will require that Julie understands how medical devices are employed in a hospital setting and considers continuity versus discontinuity with how things are currently done. The norm of clarity will require clear documentation as well as transparency about the capabilities and risks of medical devices. Attention to social norms will inform how a design team works together and how final products shape the social interactions of the medical professionals who use them. In this case study, social norms will have implications for how Julie relates with her coworkers, supervisor, employer, and customers, which will inform how she communicates the concern about the valve. The norm of stewardship concerns the cost of the design and how parts are procured but also the environmental considerations during manufacturing and when the product reaches end of life. The norm of harmony will influence the user interface design of medical devices so that they are both intuitive and pleasing to operate. The norm of justice deals with compliance with medical standards as well as efforts to make medical devices available to as many people as possible. The norm of caring may have influenced Julie's decision to work with medical devices, but it should also inform her posture toward her coworkers and customers. Finally, the norm of faithfulness deals with the reliability of the medical devices Julie designs, and it also concerns where Julie puts her ultimate trust. The Christian faith places trust in God, but also includes spiritual practices to shape a person's character and ability to discern norms in a particular circumstance.

CONCLUSION

In summary, engineering ethics often proceeds by offering ethical theories and frameworks that can be applied to case studies and then used to decide how to act in a given situation. Various professional codes of ethics can serve as helpful resources when navigating difficult situations. While it is important for engineers to engage with these professional ethical frameworks, N. T. Wright suggests that "ethics tends to provide a very restrictive view of what human life is about." He encourages us to approach "ethical questions" through the "larger category of the divine purpose for the entire human life."[29] The entirety of human life is made up of many normative aspects, as we discussed in chapter four (including cultural, social, aesthetic, justice, and faith aspects of creation), so we cannot simply reduce ethics to something we pull out when we face a dilemma. It is part and parcel of living into our callings, "improvising" faithfully as Christ's agents of reconciliation (2 Corinthians 5:18-20).

Our posture in our work is a matter of witness as well as obedience: "Let love and faithfulness never leave you; bind them around your neck, write them on the tablet of your heart. Then you will win favor and a good name in the sight of God and man" (Proverbs 3:3-4).

The reason we decided to name this chapter "Beyond Engineering Ethics" is that we want our readers to recognize that professional ethics, and indeed engineering itself, needs to be understood within the larger context of living within the biblical story. In the next chapter we will look more closely at the negative impacts of sin which occur when we choose to live within a different story.

CHAPTER SIX

MODERN TOWERS OF BABEL

Pride goes before destruction, a haughty spirit before a fall.

PROVERBS 16:18

Derek Schuurman continues as our guide in this chapter by exploring the ways sin impacts engineering, technology, and the human heart. Even so, technology remains a part of creation which, in principle, can be directed in God-honoring ways despite the possibility for sinful distortions.

On April 10, 1912, a massive new passenger liner, *RMS Titanic*, slipped out of the Southampton harbor in England to make its maiden voyage across the Atlantic. With an estimated 2,224 passengers and crew aboard, the ship headed for New York City. The *Titanic* was the largest ship of its time, and the first-class amenities included a swimming pool, gymnasium, libraries, and spacious cabins. The ship was considered by many to be "unsinkable" with its design of watertight compartments.[1] The captain of the *Titanic*, Edward Smith, had previously remarked about a similar large ship, the *Adriatic,* by saying, "I cannot imagine any condition which would cause a ship to founder. I cannot conceive of any vital disaster happening to this vessel. Modern shipbuilding has gone beyond that."[2]

Four days into the voyage, on April 14, the ship hit an iceberg in the cold, dark waters of the Atlantic. The collision caused damage along the starboard side causing several watertight compartments to begin flooding. The ship had only twenty lifeboats, far less than what was needed for

Figure 6.1. The *Titanic* sinking

all the passengers onboard. Tragically, many perished in the frigid waters of the Atlantic as the immense ship gradually sank. Even as reports from the *Titanic* began to come in, Philip Franklin, vice-president of the shipping company responsible for the *Titanic*, told reporters, "We place absolute confidence in the *Titanic*. We believe that the boat is unsinkable."[3]

In a 1991 expedition, filmmaker James Cameron captured video of the wreck of the *Titanic* showing dramatic footage of the ship's bow and ghostly images of the inside. Even as a wreck, the opulence of the *Titanic* is still visible amid the grand staircase inside the ship and the ornate artifacts strewn along the ocean floor. Nearly fifteen hundred people died, placing the *Titanic* among the largest peacetime marine disasters in history.

The inquiries into the sinking of the *Titanic* led to several conclusions: the number of lifeboats was inadequate, the captain had failed to heed ice warnings, and the ship was steaming too fast. Many of the stories that followed the disaster underscored the technical hubris that accompanied the sinking of the "unsinkable" *Titanic*. The 1997 movie *Titanic* captured this in the memorable lines exchanged soon after the ship hit the iceberg. White Star Lines Chairman J. Bruce Ismay declares, "But this ship can't sink!" to which Thomas Andrews, the architect of the *Titanic*, replies, "She's made of iron sir. I assure you she can. And she will. It is a mathematical certainty."[4]

This chapter explores some of the impacts of sin on our technology. Christians understand that all creation groans under the weight of sin. We who work in technological professions are not immune—we should not be surprised that sin also taints our dreams and the technologies we pursue. You may recall that chapter two introduced the idea that, though God's creation is good, it was corrupted by humankind's fall into sin. In chapter four we looked at norms as part of the structure of creation. In the previous chapter we explored professional ethics as understood within the larger biblical story. In this chapter we will explore the consequences of living within other stories.

THE TOWER OF BABEL

Undue pride in technology can lead to engineering disasters. It can also distort the meaning of technology, shifting its purpose from serving God to pursuing human autonomy. The Bible includes a story about human pride that ended in a failed engineering project. In Genesis 11, we read about the tower of Babel, an architectural project aimed at building a city "with a tower that reaches to the heavens" (Genesis 11:4). This project was enabled by using enhanced construction techniques, namely baking bricks and using tar for mortar.

Genesis 11:4 tells us that the motivation for building the tower was threefold. First, they wanted to make a tower that reached to the heavens, perhaps intended to serve as a bridge between humanity and the gods. Second, they wanted to make a name for themselves, an act of pride. And third, they wanted to build a tower so they would not "be scattered over the face of the whole earth." This last reason was in direct opposition to the cultural mandate of Genesis 1:28 that called humankind to "fill the earth" rather than staying in one place. Even though they were building a tower to reach the heavens, we read that

God "came down" to see the tower (Genesis 11:5). The irony in this verse underscores the insignificance of human technical ambitions in relation to God's greatness. In the end, God intervenes, the project fails, and construction is halted. As a consequence, people are scattered instead of making a name for themselves, and the name Babel is forever associated with confusion and a failed building project. The tower of Babel story aptly demonstrates the consequences of giving in to the temptation to seek self-sufficiency and autonomy apart from God.

PRIDE AND HUMAN AUTONOMY

To understand the human impulse toward pride and autonomy, let's look at the biblical story of the fall in the Garden of Eden. In Genesis 3, Adam and Eve defied God's instructions and ate the fruit from the forbidden tree of the knowledge of good and evil. Despite God's warning not to eat the fruit (Genesis 2:17), Adam and Eve fell prey to the temptation that "your eyes will be opened, and you will be like God, knowing good and evil" (Genesis 3:5). They were tempted by a desire to be like God and to pursue their own autonomy and independence. The Bible makes it clear, however, that the fall had catastrophic consequences for the whole creation. First, the fall changed humanity's relationship with God, illustrated in the way that Adam and Eve hid from God after disobeying him. Second, the fall hurt their relationships with each other. This is illustrated in the way they see themselves as naked and in how Adam shifts blame to Eve when God confronts them. Lastly, the fall's curse damaged their relationship with creation, bringing thorns and thistles that made work harder and bringing pain in childbearing.

In fact, the effects of sin go beyond individual relationships; they are comprehensive. The book of Romans tells us that "the creation was subjected to frustration" and "the whole creation has been groaning" (Romans 8:20, 22). In the words of a song by Bob Dylan, "Everything is broken." The implications of the fall extend to all human activities, including the realm of technology.

So, should we avoid technology? Is the story of the tower of Babel a warning for those who aspire to be engineers? Let's go back to Scripture. One of the first things we read after the story of the fall is how God fashioned "garments of skin" to clothe Adam and his wife (Genesis 3:21). In this act, God shows how the earth's resources can be fashioned and used, even in their fallen state. In another example, only two chapters prior to the tower of Babel story, we read about another technology project, Noah's ark. In the story of the ark, we see God through Noah employing maritime technology to save humanity and animals from the flood. Genesis 6:14-16 recounts how God himself lays out the "blueprints" for the ark, giving Noah specific dimensions, materials, and plans to follow. After the ark saves them from the flood, God renews his call for humans to be fruitful and to fill the earth (Genesis 9:1). This story seems to affirm the value of technology that is directed in obedience to God and, at the same time, affirms the cultural mandate to develop the latent potential in creation.

One might observe that these two examples are precipitated by sin. Indeed, without sin Adam and Eve would not have needed clothes and the ark would not have been necessary as a rescue from the flood. Furthermore, without sin we would not need medical technology nor military technology. Is it plausible that technology is a result of the fall? Philosopher Jacques Ellul explores this idea in an essay titled "Technique and the Opening Chapters of Genesis." For Ellul, the word *technique* represents not only technology but an entire technological way of thinking. Ellul suggests, "Thus, no matter what attitude one takes toward technique, it can only be perceived as a phenomenon of the fall; it has nothing to do with the order of creation."[5]

While it is true that certain technologies exist to address the consequences of the fall, it is not true for technology in general. I can imagine Adam and Eve, in their prefallen state, eventually forging kitchen gadgets, garden tools, furniture, and devices for making art and music. We must recall that, even before the fall, the potential for technology was part of the goodness of creation (as described in chapter two).

Technology, then, is not a *result* of the fall even though it is *impacted* by the fall.

THE TECHNOLOGICAL WORLDVIEW

To be fair, Ellul is referring not to what we normally think of as technology but to *technique*. By technique, Ellul means "the *totality of methods rationally arrived at and having absolute efficiency* (for a given stage of development) in *every* field of human activity."[6] Applying rational methods to *everything* is to treat everything like a machine. This "mechanical world picture" is one in which nature, including human nature, is conceived as an elaborate mechanism.[7] The tendency to see everything through a technical lens results in a "technological worldview," a way of seeing the world that views created reality as a technical object to be manipulated and reduces problems to technological challenges. Such a worldview, even if it is never acknowledged, can be discerned in efforts to control nearly everything through rational-technical means. These efforts toward technical control even extend to areas such as religion, art, morality, law, and politics.[8]

The philosopher of technology Lewis Mumford has traced the way technological developments over centuries were accompanied by a change in thinking and the emergence of a new way of looking at the world.[9] A posture of technological control emerged during the Enlightenment and is reflected in slogans like "Knowledge is power," a phrase which has often been attributed to Francis Bacon. A technological worldview does not view creation as something to be *stewarded* but as something to be *manipulated* for power and control.

By the late nineteenth and early twentieth centuries, this worldview was even applied to managing people. Various time and motion studies were used to find the most efficient method for a human to perform a task.[10] The techniques devised have been applied in factories and on assembly lines and, when used solely to maximize production, can reduce people to cogs in a machine. The iconic 1936 movie *Modern Times* amusingly depicts the outcomes of adopting such a worldview.

Actor Charlie Chaplin plays a hapless factory worker desperately trying to keep up with a fast-moving assembly line, eventually being swallowed into the gears of the machinery itself. Although the film is a comedy, it depicts the oppressive drudgery of industrialization. A world in which everything is seen as a machine soon becomes "fit only for machines to live in."[11] The end result is that creation's diverse physical, biological, political, cultural, social, and aesthetic realities are flattened as everything becomes an object of technological control.

C. S. Lewis compares technology viewed this way to magic. In his book *The Abolition of Man*, Lewis writes, "For magic and applied science alike the problem is how to subdue reality to the wishes of men: the solution is a technique; and both, in the practice of this technique, are ready to do things hitherto regarded as disgusting and impious." Lewis refers to this as the "magician's bargain" whereby "man surrenders object after object, and finally himself, to Nature in return for power."[12]

Ironically, seeking control through technology often leads to a loss of control. The animated Disney film *Fantasia* illustrates this beautifully. Mickey Mouse takes on the role of a sorcerer's apprentice faced with the chore of filling a cauldron with water by lugging buckets of water. In the absence of the sorcerer, Mickey puts on the sorcerer's hat and commands a broom to grab a bucket and haul the water for him. At first things go well, but as the story unfolds Mickey envisions the broom multiplying out of control until the whole room fills with water. So it is with technology—although the human ingenuity that created the technology can seem magical and amazing, it can unleash unintended consequences, even for the experienced engineer.[13] Theologian Eugene Peterson compares modern technologists to pagan magicians when the goal is to "manipulate nature for selfish benefit." Peterson writes, "The means have changed, but the spirit is the same. Metal machines and psychological methods have replaced magic potions, but the intent is still to work our will on the environment whether it involves people or place."[14]

Pursuing the magic of technology for selfish benefit has clearly led to environmental consequences. Lynn White's provocative seminal paper "The Historical Roots of Our Ecological Crisis" places the blame with Christianity for the current environmental crisis.[15] White claims that the Christian creation story replaced a sense of the sacredness of nature with an attitude of subduing nature, which in turn led to a posture of technological mastery and control. Although his critique misrepresents a true reading of Scripture, it stings because the truth is that Christians have contributed to our current environmental crisis and have not always been at the forefront of environmental stewardship. In the creation story, we read that "the LORD God took the man and put him in the Garden of Eden to work it and take care of it" (Genesis 2:15). The painful truth is that we have "worked" the creation but have not heeded the associated call to "take care of it." However, the issue is not the cultural mandate, as White suggests, but rather how this passage has been distorted by a technological mindset. A faithful biblical approach must take seriously the call for care and stewardship of God's creation in all its aspects.

While creation care should not be reduced to a technical endeavor, engineering provides a practical avenue to exercise care and stewardship. Engineers can employ sustainable engineering practices in their designs, such as lifecycle analysis, to assess the environmental impacts of products from manufacturing ("cradle") to disposal ("grave"). Engineers can also select materials and design approaches that influence and promote sustainability. Examples include RoHS (reduction of hazardous substances) certified parts in electronics, LEED (Leadership in Energy and Environmental Design) certified designs in civil engineering, and guidelines from the AIChE (American Institute of Chemical Engineers) Institute for Sustainability.[16]

Unfortunately, Christians are not immune to the temptations of the magic of technology. In his book *The Screwtape Letters*, C. S. Lewis writes about an imaginary exchange of letters between a senior demon named Screwtape and his nephew, Wormwood. In one letter, Screwtape writes, "What we want, if men become Christians at all, is to keep them in the state of mind I call 'Christianity And.' You know—Christianity and the Crisis, Christianity and the New Psychology, . . . Christianity and Vegetarianism, Christianity and Spelling Reform. . . . Substitute for the faith itself some Fashion with a Christian colouring."[17] Christian engineers need to be on guard against a "Christian coloring" that would adapt Christianity to a technological mindset. The Bible is clear that "no one can serve two masters" (Matthew 6:24). Our hope must rest in Christ alone if our technology is to be placed in service of God and neighbor. Philosopher Martin Buber warns, "Whoever knows the world as something to be utilized knows God the same way."[18]

> **TECHNICISM** is the trust in the progress of technology to solve all our problems and to bring health and material prosperity.

A "Christian coloring" can also be applied to a technological worldview. Trusting in God alongside technology is, in essence, idolatry.[19] As John Calvin put it, "Man's nature, so to speak, is a perpetual factory of idols," and technology, like anything else in creation, can be made into an idol.[20] Oftentimes technology is a "surface idol," something we use to pursue deeper idols such as power, control, or autonomy.[21] Idolizing technology in this way is sometimes referred to as *technicism*. Technicism is the trust in the progress of technology to solve all our problems and to bring health and material prosperity. Some even look to technology as a solution to the problem of death (this will be discussed in chapter eight).

It may be helpful to note here that idolatry is not simply a set of ideas or practices; idolatry profoundly misshapes us. As image bearers, we were made to image and reflect our Creator. By trusting something other than God, we begin to resemble the thing we trust. In Psalm 115, we read about idols and how "those who make them will be like them, and so will all who trust in them." In his book *We Become What We*

Worship, theologian G. K. Beale puts it this way: "What you revere you will resemble, either for ruin or for restoration."[22] C. S. Lewis offers us a helpful illustration in *The Voyage of the Dawn Treader,* in which Eustace, a selfish young boy, stumbles across a cave with a large treasure belonging to a dragon. Eustace falls asleep on the treasure and dreams of power and riches and awakes to find he has turned into a dragon! Lewis writes, "Sleeping on a dragon's hoard with greedy, dragonish thoughts in his heart, he had become a dragon himself."[23] Likewise, if one reveres technology, one will begin to resemble it, becoming more cold, calculating, efficient, and pragmatic. For example, consider a computer scientist who is so enchanted by algorithmic thinking that he begins to apply it in many other areas of life, such as relationships and career decisions.[24] When technology is revered at a societal level, it can lead to "the submission of all forms of cultural life to the sovereignty of technique and technology," something Neil Postman referred to as *technopoly.*[25]

How does technology actually (mis)shape us? Technology shapes our experiences and how we see the world, whether we are aware of it or not.[26] Recall from chapter three that technological design is not neutral, and built-in biases and preferences make some things easier and other things more difficult. Our everyday technological habits and rituals are like "liturgies" that shape our hearts and lives over time. A "liturgical audit" of our technology usage can help to ensure we are cultivating healthy rituals and rhythms.[27]

Technology often nudges users to establish particular practices and habits, impacting behavior by making some things easier and other things more difficult. For example, digital video streaming services which automatically play the next episode can encourage viewers to watch more, and speed bumps in roads can influence drivers to travel more slowly. The question engineers can ask is, What habits do the devices we design nudge users toward? Do our designs encourage more distracted living and increased consumption, or do they help promote healthy social habits and contentment? Of course, raising these issues in a secular, corporate setting would require translating the language of "liturgies" into words like "digital well-being," "customer experience," or "design for social good," but a sensitivity to cultivating helpful habits and liturgies can inform our engineering work for the well-being of all users.

It is not only technology that shapes us. As engineers, we should also be aware of the habits and practices we foster in our professional lives. Our work practices are formative and are directed toward certain visions of the good life. For example, long hours spent at work fueled by dreams of technical and business success will gradually shape us. Our education also shapes us. Since engineering education usually focuses largely on technical skills, engineers may be inclined to approach most problems as technical problems amenable to technical solutions. Although rarely acknowledged, this tendency can foster an undue confidence and trust in technology. We can also be shaped by a psychological phenomenon known as "groupthink" in which one sets aside personal beliefs in order to maintain consensus within a group. Christians serving in the world of technology need to resist such influences which can seep into their hearts and minds. To borrow a phrase from media ecologist Neil Postman, we must strive to become "loving resistance fighters."[28]

SIN AND CONSEQUENCES

There is a familiar saying in technology circles that "to err is human, to really foul up requires a computer." The fact is that technology has the power to amplify the consequences of sin. Sin's impact on technology and engineering can have devastating and widespread consequences for many others. Because of this, the National Society of Professional Engineers code of ethics begins with the requirement to "hold paramount the safety, health, and welfare of the public." This call to "hold paramount" the welfare of the public recognizes that a technical artifact is not just bits, bytes, wires, gears, and concrete, but it includes social, cultural, economic, justice, aesthetic, and ethical considerations (as explored in chapter four). The following stories demonstrate how technology can amplify sin and destruction.

In 2016, malicious software was discovered that targeted "internet of things" (IoT) consumer devices such as home routers, security cameras, and home appliances.[29] The malware spread by scanning for insecure IoT devices using default usernames and passwords and then infecting them. The devices were turned into a "botnet" and harnessed to perform large scale network attacks, interfering with web services using a "distributed denial of service" (DDOS) attack. The source code for the malware was posted on internet forums and has since been adapted and applied to other attacks. Creators of malware intentionally create tools that bring harm to others, motivated by sins such as greed, pride, and enmity.

Roger Forsgren, in an article titled "The Architecture of Evil," writes about how Hitler's chief architect, Albert Speer, used "his brilliant technical expertise and talents to enable the war efforts of the most evil regime in history, allowing it to murder millions of human beings."[30] After the Nazi defeat and during his trial at Nuremberg, Speer argued that "he was not personally involved in the atrocities committed by Hitler and his henchmen. . . . He was simply a technical person, an architect, and was unaware of the worst crimes being committed by the Nazi regime until it was almost too late."[31] Despite Speer's excuses, he was convicted and sentenced to twenty years in prison. Although this story is an extreme example, it is a reminder that engineers bear responsibility for the technical work that they do. Speer himself frankly points to certain habits and practices that led to a disregard for human life: "My obsessional fixation on production and output statistics, blurred all considerations and feelings of humanity. An American historian has said of me that I loved machines more than people. He is not wrong."[32]

In chapter one, we suggested that dreams inspire the pursuit of technology, but it is important to note now that we do not always dream good dreams. Our technological dreams must be guided by norms if we are to amplify the good possibilities in creation. Roger Forsgren concludes that "engineers . . . have made man fly; and architects have constructed buildings that reach thousands of feet into the sky. But these same technical gifts alone, in the absence of a sense of morality and a capacity for critical thought and judgment, can also make reality of nightmares."[33] Troubles arise when we allow sin to shape our dreams or when we fixate so much on technical aspects that we develop a sort of tunnel vision that ignores wider considerations. In the words of economist Bob Goudzwaard, "If man and society ignore genuine norms, . . . they are bound to experience the destructive effects of such neglect. This is not, therefore, a mysterious *fate* which strikes us; rather it is a *judgment* which men and society bring upon themselves. . . . Genuine laws or norms are pointers that guide us along safe and passable roads. Apart from norms, our paths run amok."[34]

FINITENESS VERSUS FALLENNESS

We might find ourselves asking whether every engineering failure is due to sin. Consider the following true stories.

In 1980, an oil company was doing some exploratory drilling in Lake Peigneur in Louisiana. They calculated where to drill based on reference points. The drilling commenced, but at a depth of 370 meters something went wrong. Below a portion of the lake was a salt mine which had operated since 1920. The location for the drilling had been carefully selected to avoid the mine below, but unfortunately a mistake had been made in the calculations. The modest drill hole swelled into a large whirlpool roughly four hundred meters in diameter as the entire lake (all ten million cubic meters of water or so) emptied into the salt mines below like a large bathtub with its plug pulled out. Thankfully, the drillers on the drilling platform and the miners below all escaped unharmed, as did one hapless fisherman who happened to be fishing on the lake at the time.[35]

Another oft-cited engineering failure example is the Tacoma Narrows bridge which opened in July 1940 (see fig. 4.3). Soon after it was constructed, observers noted that the suspension bridge would sometimes move up and down in the wind, giving it the nickname "Galloping Gertie." A few months later, in November 1940, the bridge began to oscillate wildly until it dramatically collapsed into the strait below.[36] Thankfully, no people were hurt, although a dog was killed. It was not until later that engineers attributed the cause to something they called aeroelastic flutter. Discovering this condition led engineers to reinforce existing bridges, and it became part of the standard design considerations in future bridge projects.

A third example occurred on June 6, 1996, with the first launch of the Ariane 5 rocket. The rocket veered off course and was destroyed forty seconds after liftoff. The cause was later traced to a part of the software where numbers representing velocity caused an "overflow error." The failure occurred even though the code had been successfully used on a previous rocket, the Ariane 4. The failure destroyed both the rocket and a satellite payload costing about $500 million.[37]

These engineering failures, which seem to be related to human finitude, remind us of our limitations and creatureliness. We are not all-knowing and cannot predict all of the consequences of our design decisions, especially when they uncover previously unknown failure modes. What's more, we also make mistakes in calculations and assumptions. Human finitude is not a sin, but humility and prudence are important in the face of finiteness and uncertainty. Furthermore, once a new failure mode is uncovered, we have a responsibility to address it as engineers.

Sometimes the difference between finiteness and fallenness is difficult to discern. For example, is a software bug due to fallenness or finiteness? In a perfect world, could humans write bug-free code? Certainly, some bugs result from lack of care or sloppiness, but could it be that the process of debugging is an intrinsic aspect of writing complex code?[38] One might argue in hindsight that engineers should have explored the motion of the Tacoma Narrows bridge earlier, or that the drilling engineers on Lake Peigneur should have done more to check their calculations, or that more software testing should have been performed on the Ariane 5. Especially in engineering projects where human safety might be put at risk, engineers have a professional and legal duty to exercise a high standard of care.

It is helpful to distinguish between our finiteness and fallenness, between creatureliness and sinfulness. Making this distinction is important because we are called to accept the former while fighting the latter. In fact, not accepting our limitations as creatures leaves us open to the sin of pride, leading to negligence due to overconfidence or working in areas outside one's competence.

As engineers, we need to be aware of both our sin and limitations, employing best practices to minimize the impact and severity of flaws. When choosing between design alternatives, we ought to use caution when considering untried methods (sometimes referred to as the "precautionary principle"). Employing safety factors in designs, performing rigorous testing, and engaging in peer reviews are crucial parts of the design process—as is a posture of humility. Such a posture can lead us to design with more caution, particularly in safety-critical systems, taking steps to build in redundant design features that help limit damage when unexpected things happen.

TECHNOLOGY STRUCTURE AND DIRECTION

We can find stories warning about the dangers of technology that go back thousands of years. One ancient Greek myth describes Prometheus who defied the gods by stealing fire and bringing it to humanity (an act for which he was later punished by the gods). Fire provides warmth and the ability to power engines, but it can also cause immense destruction and harm. The Prometheus myth informs the subtitle of Mary Shelley's famous nineteenth-century novel *Frankenstein or, the Modern Prometheus.* Shelley's story describes a young scientist who discovers a technique to create life and proceeds to build a monster that eventually turns on him. The Frankenstein narrative is one that is repeated frequently in science fiction stories, often taking the form of rogue robots or artificially intelligent computers that threaten to destroy humanity. Technology can bring life and flourishing, but both fiction and reality remind us that it can also bring chaos and destruction.

Thinking about the allure of technology, the pitfalls of technicism, and the fears of technology run amok can make one ask whether Christians ought to distance themselves from the perilous world of technology. When considering such a question, it is helpful to think in terms of *structure* and *direction*. Theologian Al Wolters describes *structure* as "the order of creation, . . . the constant creational constitution of any thing, what makes it the thing or entity that it is" and the concept of *direction* as the way anything in creation "can be directed toward God or away from God—that is, directed in obedience or disobedience to his law."[39] This direction "applies not only to human beings, but also to such cultural phenomena as technology, art, and scholarship, to such societal institutions as labor unions, schools, and corporations."[40] Wolters makes a key point: we ought not to reject technology or any other possibilities in creation but discern how they are oriented—either toward obedience or toward disobedience to God. Examples of direction for technology might include green technologies versus machines that produce unmitigated hazardous waste, or software designed to assist with medical diagnoses versus software designed to be addictive in order to monetize the attention of its users.

The concept of direction helps us understand that sin does not create anything on its own—sin is like a virus that attaches itself to creation like a parasite. It may sound unusual, but even something like internet pornography is only possible because of God's creation: he created the possibility for computer technology as well as human sexuality, but internet pornography is taking God's good creational gifts and twisting and perverting them in a way they were never intended to be.[41] We can direct technology in obedience to God or toward more disobedient uses. Asking whether technology is good or bad presents a false dichotomy. Technology is, in fact, part of God's good creation. The question to ask is, *To what direction is it pointed?*

Examples of structure and direction are found throughout Scripture. The book of Genesis opens with the familiar creation story and includes a curious parenthetical note: "The gold of that land is good; aromatic resin and onyx are also there" (Genesis 2:12). These raw materials and the latent potential that they possess are mentioned as part of God's good creation. However, later in Scripture we read that gold is fashioned into a calf just a few chapters after God commanded them not to make gods of gold or silver (Exodus 20:23). In Daniel 3, we read about King Nebuchadnezzar's image of gold that people were required to worship. This story also illustrates how created materials can be directed, using technical and artistic skills, toward idolatry rather than God. In contrast to this, we read about the first person the Bible records as being filled with the Spirit, someone who was not a king or prophet or priest, but a craftsman. Bezalel was filled "with the Spirit of God, with wisdom, with understanding, with knowledge and with all kinds of skills—to make artistic designs for work in gold, silver and bronze, to cut and set stones, to work in wood and to engage in all kinds of artistic crafts" (Exodus 35:31-33). Here gold and other raw materials were skillfully crafted and put in the service of God in the construction of the temple. Later, in Matthew 2:11, we read that the wise men brought gold as a gift to Christ. It is interesting to note that the materials gold, onyx, and pearls, which appear in Genesis 2 (the second chapter of the Bible), reappear in Revelation 21 (the second-to-last chapter of the Bible). Revelation 21:18-21 describes the holy city, the new Jerusalem, as a "city of pure gold" and one decorated with precious stones, including onyx.[42] The heavenly city stands in stark contrast to other biblical cities, like Babylon and those built by Cain and Nimrod.

In summary, what distinguishes technological artifacts in these stories is neither the materials used nor the technical skills employed, but how they are *directed*. Our cultural artifacts can be directed in obedience to God or in disobedience to him. God does not oppose technology, but he does require that it honors him and builds his kingdom. We can use the same technical skills and materials to build temples and tools in the service of God or to build golden calves and towers of Babel. When directed properly, engineering can indeed be a gift that allows us to show love for our neighbor and to bring glory to God.

CONCLUDING THOUGHTS

In the end, we must attend to our hearts. Any efforts to build modern-day towers of Babel to bridge heaven and earth by our own efforts will be futile. Thanks be to God, sin does not have the last word and Christ himself has provided the bridge between heaven and earth. As Christian engineers, we are called to participate in Christ's work as agents of reconciliation (2 Corinthians 5:19), directing technology toward loving our neighbors and bringing glory to God. The next two chapters will provide a chronological perspective to help us discern the directions in technology, by looking back historically and then looking forward to the future.

CHAPTER SEVEN

MOVING FORWARD BY LOOKING BACK

It is a good rule after reading a new book, never to allow yourself another new one till you have read an old one in between.[1]

C. S. LEWIS

Engineers take joy in invention, dreaming of novel technologies. However, this forward-looking perspective sometimes blinds us to historical lessons and context. In this chapter, Ethan Brue leads us on a fascinating field trip through past industries. Reading this chapter will cultivate an important skill for technology developers: to continually "zoom out" to both see and imagine the big picture in historical context and the overarching biblical narrative when immersed in the daily details of a complex project.

Imagine for a moment that time travel is so routine that you hop into your ATV (All Times Vehicle) with no particular destination in mind on an uneventful Saturday afternoon. Closing your eyes, you flip the dashboard controls, experience an intense rush of acceleration, and as the dancing vapors clear from your windshield you catch a fleeting glimpse of the Statue of Liberty to the right and historic St. Patrick's Cathedral to the left. Screeching to a halt just past Fifty-Second Street on Broadway, the hipster-style sign in front of a thriving business reads, "New York Electric Vehicle Company." You assume that this must be a Tesla spinoff that has morphed from its California roots and expanded coast-to-coast. As a resident of middle-America during the early years of the plug-in electric automobile, you guess that you have landed in the mid to late twenty-first century.

You have arrived in an era in which the electric vehicle is not only a promising technology but a thriving global business, ripe with future potential. The Electric Vehicle Company in front of you is the largest vehicle manufacturer and operator of motor vehicles in the United States, the first corporation to transition from custom product manufacturing to standardization, and a company whose sales network stretches from San Francisco to Mexico City to Paris.[2] While it may come as a surprise to you, you have not landed in the future but in the past. You are now parked in the second half of the nineteenth century.

This historical excursion should enable us to see that today's problems have originated many years before we first imagined solutions. We create technological solutions, but we also invent many of the perceived needs and desires as well. This principle is best explored by example, using one particularly interesting and complex technology, the electric vehicle. This technical vignette demonstrates how historians have an unfair advantage. They have access to the closing chapter of stories that were previously inaccessible. Historians of technology would know that within a few decades, those early electric vehicles disappeared and were seldom remembered. The story is a testament to the mortality of humans and their engineering endeavors. Historians also carry an inevitable handicap. Being culturally bound in a specific time and place, historians of technology all too often conclude that the "winning technology was the best because it won, and it won because it was the best."[3] We are tempted to assume that what we build and create is progress because it occurs in the future relative to the past.

Nobody needs to time travel to acquire historical humility. Studying history should do this for us. We tend to see electric vehicle development as progress, but history forces us to qualify or give nuance to this conclusion. To explain why a particular culture decides to take what we now consider "the future," bury it in "the past," and then undertake technological grave-digging to reintroduce it multiple generations later is an ongoing debate. The modern electric vehicle industry (like other technology marketed today) claims no ancestry and continues to promote its product as novel without ever implying that it is engaging a technological solution that was more than once rejected. To help us understand this early rejection of "the future" in automobile development, we will first consider some possible theses that could explain the failure of the electric automobile and the triumph of the internal combustion vehicle in American culture.

BACK TO THE FUTURE: FROM BICYCLES TO ELECTRIC VEHICLES

As we discussed in the first chapter, dreams matter. The American dream of *automobility* represented much more than the hope of replacing the horse as a means of land transportation.[4] Our social imagination for automobility was not driven by the need for mechanized travel from New York to Chicago or any other important city center. Already by 1899, the railroad and steam locomotive filled this need. The dream of automobility was a desire to escape ordinary life whenever and wherever a person wanted to do so.

Automobility as a cultural phenomenon first went viral not as a car but as a bike. This late nineteenth-century bicycle obsession inspired the "safety bike," a chain-driven device with equal size wheels, that exploded onto the market in the 1890s. Long before Henry Ford marketed the American dream of a vehicle for every household, bicycle manufacturers had convinced Americans—not just the wealthy—that they needed a safety bike for every person. This egalitarian dream was rooted in the human need for play. Bicycle technology was for evenings and weekends, allowing people to escape with friends or family to the country, lakefront, or regional wilderness. Along with this bike craze came hundreds of new cycling clubs and associated periodicals for bike owners, promoting racing and recreation.[5] It is estimated that by 1895 four million Americans owned bicycles and nearly as many rented them.[6] It is no coincidence that by 1897, when the bike boom began to fade, the League of American Wheelman did not close its doors. They changed their name and their technology but not the dream. They became the American Road Makers. It is not uncommon for a dream to outlive its technology.

This is the cultural climate in which the early electric automobile was taking shape. While many reasons are given for the early demise of the electric vehicle, this overarching dream of automobility colors them all and gives us a case study both on how we never escape our past and how our beliefs regarding progress reveal our religious orientation. Let's look at a few of the possible explanations for the failure of the early electric vehicle.

Thesis 1: The early electric vehicle failure was the result of heavy maintenance, reliability, and weight challenges intrinsic to battery technology available at that time. Early electric automobiles were plagued with maintenance challenges. Lead-acid batteries have a long history of excessive maintenance, extending well into the late twentieth century. The heaviness of the electric vehicle was the direct result of its batteries. This cumbersome load, combined with minimally maintained roads, arguably led to increased failures in the body, frame, wheels, and suspension. Inadequate infrastructure and frequent maintenance aside, the energy density (the energy to weight ratio) of the charged battery in an electric vehicle compared very unfavorably to the energy density of the full gas tank in the internal combustion vehicle.

To this day, electric-powered vehicles cannot match their gasoline powered counterparts when comparing stored energy per kilogram of mass. It is important to recognize that this comparison assumes that the goal was to escape city life via long-distance transportation technology independent of train routes. If this was not the vision, neither the weight, battery density, nor the condition of rural roads would have been a serious impediment for the early electric automobile. Urban commuting and a fleet of trucks to distribute goods regionally demand a different set of evaluation metrics. Technology built for local traffic on well-maintained city infrastructure will not yield the same reliability when used for wilderness travel or overused urban arteries. Any measure of reliability needs a common context for meaningful assessment. A focus on the lack of reliability of the early electric vehicle often leads us to conclude that the internal combustion engine was proven or maintenance-free technology at this point in history. But gasoline driven vehicles had their own assortment of start-up challenges. Neither should we overlook an established infrastructure of stables, veterinarians, blacksmiths, feed producers, and rendering facilities that masked the maintenance and reliability issues associated with the horse. For all technology options in the automobility mix at the turn of the nineteenth century, reliability remained a goal, not a reality.

Thesis 2: The early failure of the electric vehicle was the result of inadequate infrastructure for electric vehicles. The electric vehicle history highlights a common social debate that has accompanied the development of a wide range of technologies throughout history. Cultural-formative norms are often characterized by questions of opposites. Three sets of opposites were central to the story of the electric vehicle: integration versus differentiation, large scale versus small scale, and centralized versus decentralized. Early electric vehicles were being developed while centralized electric power generation was taking root in America's largest cities. This meant that the local infrastructure for electricity generation was well established in cities across America long before the infrastructure for liquid fuel (a.k.a. gas stations) for automobility. Central generation stations had plenty of capacity to charge large fleets of electrical vehicles overnight and in the evenings when their primary industrial customers did not need power.

Despite this natural affinity between the developing electric vehicle and central generation units, there was also cultural resistance growing from outside the technical affinities. Power generation in the early twentieth century was local. Power plants were usually located next to industrial and commercial users in city centers. The cultural vision of automobility, to provide an escape, was antithetical to everything the city and its industrial centers represented. The dream of automobility defined Americans first as individuals, and only second as a community. This automobility ideal led people to embrace power generation that favored decentralization (generation on the vehicle) over centralization (generation on location). Automobility favored a distributed rather than concentrated infrastructure. This change in thinking also expressed itself in efforts to move toward technological development on a large scale (nationally) rather than small scale (locally), and favor differentiation (individual) over integration (communal). The result was that the automobility dream pushed us toward the intrinsic values of the internal combustion vehicle.

Thesis 3: The early electric vehicle failed because electrically driven commuter transportation was not practical. Ironically, it may have been the practicality of electrification that contributed to the downfall of the electric vehicle. We see an early glimmer of the success for the electric automobile in the emergence of fleet vehicles at central stations and for regional product distribution.[7] In addition, the usefulness of electricity for mass transportation had already been realized in horseless trolley technology. Electric traction transportation (i.e., streetcars) was a booming industry by 1902. Within ten years of its arrival in cities, Americans took advantage of electric trolley technology 4.8 billion times, in an era that boasted only a few thousand automobiles.[8]

Electric transportation was both proven and practical. In fact, it was so successful that it compounded the already crammed and stifling experience of using the mixed mass transit systems of the city, intensifying the dream of automobility. The dream of automobility was for escape, not utility. Cars were the new bikes long before people imagined trucks or a larger transit system complete with interstate highways. The practicality of the internal combustion automobile emerged only later as a side effect. The early electric vehicle was perceived as practical and not recreational. It could thrive only as a technology tethered to the immediate locale and central generation infrastructure. It was not an avenue toward urban escape and wilderness adventure.

Thesis 4: The early electric vehicle failed because it did not effectively market itself to all consumers. It is a common practice in our contemporary world to market shoes, Bibles, bicycles, sunglasses, and a host of other products specifically to male and female consumers. It may be less so with automobiles, but this was not always the case. At the turn of the century automobility came with a gender. As Virginia Scharff describes,

> [Automobile] manufacturers tended to associate the qualities of comfort, convenience, and aesthetic appeal with women, while linking power, range, economy, and thrift with men. Women were presumed to be too weak, timid, and fastidious to want to drive noisy, smelly gasoline powered cars. Thus at first, manufacturers influenced by Victorian notions of masculinity and femininity, devised a kind of "separate spheres" ideology about automobiles: gas cars were for men, electric cars were for women.[9]

The limited range of the battery was another reason to align the electric vehicle with women, since their assumed domestic responsibilities would keep them close to home, well within the range of the best electric vehicle battery. Marketing to women was done within a social context in which some questioned whether women ought to drive automobiles at all.

While it is likely that marketers carried with them a mix of progressive and discriminatory attitudes, electric vehicle manufacturers were trying to do more than capture an untapped niche market. They were hoping to expand their market to both male and female consumers by challenging the status quo of who could be included in automobility. They argued that since the electric vehicle was easy enough for a woman to drive, comfortable enough for a woman to love, and reliable enough for a woman to trust, everyone should prefer electric automobility.

The strategy seemed to have backfired. Remember, the original dream for automobility was driven by a desire for adventure (touring), risk taking (racing), and escape (conquest), something that aligned more with the era's assumptions about masculinity. And since those who did most of the buying and designing at that time were male, it is not surprising that this attempt to reach a broader market share actually resulted in a smaller market. With marketing like this, the chances of a culturally attuned male purchasing an electric vehicle were about as likely as him investing in a hoop skirt. Technologies fail, not only when they don't perform adequately, but also when they do not resonate with the cultural imagination of society. It is not just the *what* of the dream that matters, but the *who* as well. However well-intentioned a progressive agenda might be, progress for some is not necessarily progress for all.

Thesis 5: The early electric vehicle failed because people were unaware of the environmental consequences of the internal combustion engine and the environmental benefits of the electric vehicle. One could assert that if there was ever a time in the history of America in which environmental and conservation conversation took center stage, it was between 1890 and 1920. This was an era of progressive public policy during which the National Park System, the National Park Service, the US Forest Service, and the Sierra Club were established. Prominent individuals such as Theodore Roosevelt, John Muir, and Gifford Pinchot played influential roles in the nation, prompting a debate on how to rationally conserve, preserve, use, and develop natural resources.[10] American automobility was in its infancy, and the development of the electric vehicle was as much a response to environmental issues as it was a casualty of them.

Environmental challenges were both local and national in the late nineteenth century. Within cities, environmental problems intensified as a diversity of new technologies merged into the already heavy flow of traffic. Pedestrians, horse-driven carts, streetcars, and trolleys and their electric counterparts bottlenecked mass transit and fueled urban

One of Our Nine Rare Creations
For 1912

Figure 7.1. 1912 Detroit electric car advertisement

congestion, leading to odors, noise, and disease. In New York City during this era of mixed transportation technology, some have estimated that a single day could see 2.5 million pounds of manure and 60,000 gallons of urine on the streets, not to mention the 15,000 dead horses per year that quickly added to the sensory-laden pollution experience.[11] These particulate and gaseous emissions, more overt and obvious than the pollutants yet to come, dominated the attention of engineers and city planners.

Automobility seemed to promise a resolution by offering flexible routes and scheduling, while at the same time expanding the range of where people could work and live. It was seen as a solution to the pollution problems caused by social concentration and the fixed route systems of city centers. In hindsight, the unabated exhaust and fumes of the internal combustion engine would make it seem that the electric vehicle should have come out the winner, but this was not as apparent to the citizens of America as they first rolled into the twentieth century.

The early gasoline driven automobile was an intentional response to environmental challenges of its day. In contrast, those who developed the electric vehicle in the early years had little or no sense of it being "clean and green," as many think of "clean and green" today. Electricity in its early days was predominantly coal energy, generated in central stations located right in the middle of city life. A broader perception of electric vehicles as "zero-emission" transportation did not emerge until later in the twentieth century as these large centralized power generation facilities faded from their communal presence and were often located farther from residential population centers. Knowing what we know today about "carbon-intensive" technology, the early electric automobile would not have won out against its internal combustion counterpart, which can use biofuels and hydrocarbon fuels with a lower carbon content at a higher efficiency. But this particular environmental criteria was not in the picture. The concern was social congestion and thoughtless consumption of natural resources.

The technology that best fulfilled the dream of a "cleaner-greener" world between 1890 and 1920 was the internal combustion automobile. It seemed to address the twofold tension of the early conservation movement, which was both preservation for the sake of human reflection, awe, and transcendence, and conservation for the sake of use. Both coal-generated electricity and petroleum (or biofuel) driven engines could make good on the promise to conserve national forests and preserve our nation's wilderness areas, but only the internal combustion automobile had the freedom and range to bring us "out there and away from" to "enjoy nature." Ironically, while we may have a different sense of what comprises an environmental crisis today as compared to then, the dream of automobility has stayed surprisingly intact.

Thesis 6: The early electric automobile failed as a result of failed business practices. The early history of the electric automobile is a testament to an ideal technical marriage between power producers, electric vehicle manufacturers, and local transportation companies. For a variety of nontechnical reasons, however, the relationship proved to be a rocky one. Electric vehicle manufacturers, owners of electric vehicle fleets, and the electric power producers struggled to work together; no group felt that the other group was appropriately carrying their fair share of electric vehicle development or promotion. Standardization was desperately needed but could not be agreed upon, and the partners often competed for profits rather than cooperating for success. The electric vehicle was caught in a dysfunctional family of industries that needed each other but couldn't get along.

LESSONS LEARNED BY LOOKING BACK

This brief history does not exhaust the explanations for the short-lived success of the early electric vehicle. Other stories point to greedy practices by electric vehicle manufacturers, central station owners, and fleet operators. We could lay part of the blame on the incompetence of the electric vehicle association or misinformation put out by internal combustion automobility advocates. We could describe the influence of railroad trusts or oil trusts that either directly or indirectly favored one flavor of transportation over the other. We could trace public policy debates in regard to natural resource preservation versus conservation, urban centralization versus decentralization, public versus private, regulation versus deregulation, and labor unions versus manufacturers to help us understand the path we took toward nonelectric automobility. We could lump it all together and explain it away as the natural working of a free market. We can't, however, ignore the social, political, and cultural constraints that seem to have rendered the enterprise more captive than free. Through it all, sin coupled with our worst intentions and our best intentions limited by our human fallibility brought us to the technological road we chose and the road we left behind.

Today's electric vehicle technology is different from that of the early 1900s, but its recent success has more to do with the entrenched internal combustion engine culture rather than radical design innovations. The growing popularity of the contemporary electric automobile comes not from embracing the electric vehicle's intrinsic strength, but from accepting the internal combustion engine culture. The contemporary electric automobile has focused all efforts on extending its range and expanding the cross-country infrastructure needed to use it. It has not tried to realign the electric automobile with its inherent fitness for regional transportation and use of localized power generation. The future of the electric automobile depends on whether it can satisfy the American ideal of automobility or whether this ideal will be replaced by one that values social and environmental ideals different from those set in motion in 1899.

Looking at the history of the electric vehicle gives a helpful case study for illustrating that technological progress is only partially determined by technical barriers and opportunities. Social, ethical, economic, psychological, and lingual factors all shape our dreams for technology. It would be easy to conclude that the electrical vehicle was simply "bad technology," and that "bad technology" always loses. It is easy to adopt arbitrary measurements of progress without recognizing that our culture creates the criteria for progress and that technology is then measured by these criteria. As David Kirsch so aptly states about early automobility, "technological superiority was ultimately located in the minds and hearts of engineers, consumers, and drivers, not programmed inexorably into the chemical bonds of refined petroleum."[12]

Such an elusive and ever-changing definition of progress may leave us unsettled and confused, wondering if it is possible to discern a right path with regard to technology. Samuel Florman suggests that this confusion illustrates the essential role that the study of history plays in the education of engineers and those who shape technology. He believes that most engineers "know a great deal about restraint and cooperation . . . [and are] logical, sober, and well meaning."[13] He assumes that most of them are likely the type that make good citizens. In asserting the importance of history, he believes the humanities will certainly make the engineer a better person, but not simply by making the engineer more ethical. In fact, he believes that the humanities will more likely rob them of their innocence, stain their sober character, and in a sense make them "less moral—or at least less naïve."[14] Florman suggests that this is a good thing, since it will encourage engineers to forever "take into account the imperfections and absurdities" of broken humanity.[15] If undertaken in humility, writing, reading, and remembering our history is a first step toward wisdom. Florman's admonition is similar to the biblical imperative to remember our story. The liturgies of remembrance in Scripture do not ask us to assess history from our point of view, but they do ask us to step into the story itself, to mourn brokenness and confusion, celebrate joy and hope, and ultimately see the sovereign work of God in and through it all.

Although it is somewhat common to speak of the "history of technology," technological history is simply human history. Technology began at the beginning of time. Biblically speaking, human history begins when the Creator breathes life into people and calls humanity into its proper relationships with God, each other, and creation. This story introduces a future and a past to humanity. We create culture. We create community. We create history. The term *prehistorical* is really a misnomer, based on an arbitrary technological milestone: writing. This is not surprising, because this definition of history emerged from a culture that idolized typographic culture and its accompanying rational techniques of knowing, thinking, and doing. This is the same culture that decided to classify us as *homo sapiens* (human the thinker). Even before writing, people told their stories to one another. Stories simply require communities who create the time and space to reenact, sing, tell, remember, and listen. Prewriting cultures developed many technical means of storytelling through instruments, clothing, utensils, paint, calendars, and stage. Storytelling is basic to living fully as humans.

Finding a specific starting point within human history for technology is as difficult as determining when technology moved to high-tech from low-tech, because such an assessment depends on our criteria. For example, in 2004, just over one hundred years after the genesis of the Ford Motor company, *Popular Science* magazine ran an article titled "Is This the Golden Age of Driving?"[16] A graphic within the article asks the rhetorical question, "What hath technology wrought?" The heading introduces the contestants as "The Original SUV vs. the Ultimate SUV." However, there is no indication which competitor was considered original and which was considered ultimate. The article provided nothing more than the comparative data. The rational reader was expected to arrive at a self-evident conclusion based on the facts alone.

What has technology wrought? We might conclude a century of automobile regression by interpreting the data provided as shown in table 7.1. While this may not have been the intended conclusion, it is not difficult to demonstrate that the quintessential SUV on the market nearly one hundred years after the inception of the Ford Motor company is inferior to the Model T. We need only compare the fuel efficiency, the safety and practicality for residential use, the consumer cost, the ground clearance, the built-in features that limit power usage, the excess weight, and the elimination of agonizing decision-making that accompanies a new car purchase.

Table 7.1. Comparison of selected criteria for Ford Model T and Hummer H2

HUMMER H2 (EARLY 2000s)	CRITERIA	MODEL T (EARLY 1900s)
12 mpg	Fuel efficiency	20 mpg
110 mph	Speed optimization*	40 mph
$49,395	Consumer cost (in 2004 dollars)	$10,185
9.7 inches	Ground clearance	10 inches
325 hp max	Rate of energy consumption	20 hp max
6400 lb	Weight	1200 lb
10+ options/choices	Purchasing ease	No options/choices

*A comparison of which model has optimized its top speed closest to the typical legal speeds for local transportation in most municipal neighborhoods.

Is this a fair comparison? Within the limited boundaries of the chosen rules, yes. In the context of the many tacit assumptions of automobility, no. The point is not to convince you that the Model T is the better vehicle, but rather to reveal that our cultural measures of progress are neither uniform, clear-cut, nor unchanging. The exercise tells us far more about *Popular Science* magazine and its traditional adoption of certain pillars of the automobility culture than it does about the specific technologies in question.

If we make the rules, it is easy to win the progress game. To craft a timeline, you need to determine what counts and what doesn't count as history worth recording. Most technology histories design a narrative based on scale, productivity, power, or economic volume. In such studies, we marvel at the precision of massive pyramids; we admire the scale of Roman roads, architecture, and aqueducts; we mourn the alleged decline of civilization through the volatile Middle Ages; and we praise the rebirth of classical traditions into the era of guns, ships, and imperial empires that culminates in "industrial revolutions" exploding across the world. All of this is rooted in the ability to scientifically model and technically control our environment. So even though this narrative tells a story, it is substantially incomplete. We are conditioned in our Western tradition to view history materially (economic, geographic, and technological) and chronologically, and we assume that all metrics of progress are measured in matter or time, with "more" or "newer" being most important. This is not universal, however. All cultures have fascinating technological stories to tell, and many measure progress differently.

A biblically guided view of progress can be described as having two broad metrics. First, biblical "progress" reflects the intended wholeness of the good creation; and second, biblical "progress" seeks the fullness of the renewed creation. The framework of normative principles outlined in chapter four can help us do responsible engineering by highlighting the integral coherence of creation. This framework points us to questions that help us apply biblically guided norms to a specific design in a particular time and place—it helps us design an artifact or process in ways that recognize the wholeness of created reality and avoid approaches that are reductionistic.

Beyond a particular design, it is also crucial to examine what it means to live, work, play, and serve fully before the face of God both now and into eternity. Getting these nontechnical elements right can often determine the success or failure of a technology and be more important than any particular design decision. Responsible technology requires an eye to both wholeness and fullness.

MAPPING A FUTURE TECHNOLOGY USING DESIGN GUIDES

Let's see how these design guides from chapter four help us think about designing an electric vehicle battery system. We might begin by calculating the total power needed and the energy storage required. This satisfies the quantitative laws that must be followed for a design to work. As we think about the energy density of the battery system, we ask spatial questions that help us determine the ideal weight and maximum volume required for practical vehicle use. The predictive models we develop, based on physical and kinematic laws for energy, offer clues to the boundaries for performance based on electron flow and the chemistry that describes ion interaction in a single battery cell. The laws for the quantitative, spatial, kinematic, and physical apply as we build our battery cells in series, to increase system voltage. These laws are bound by the limits of creation and the limits of our knowledge of creation. But while increasing the electric potential increases our energy delivery capacity, it also increases the potential threat to humans and other living creatures, as the risk of electric shock or burn increases. As these ethical considerations come into play, considerations that relate to biology and sensitivity, we are confronted with questions of normativity and responsibility. The analytical norms are engaged as we account for the multiple design options available to meet our objectives. The cultural and clarity norms guide us as we research the many battery attempts of the past, building on the successes while avoiding the failures of past designs. This means documenting work effectively so that, as we benefit from our historical knowledge of how creation works, others can benefit from ours. Developing the battery in isolation is difficult if not impossible. Being able to depend on fabricators, technicians, and commercially available hardware presupposes that you have established good social relationships and partnerships. Throughout the whole design process you make choices about stewardship. How much time and money can you afford to invest? At what point do the minimal gains in performance no longer justify the cost of additional time and improved materials? What will the lifecycle be for this battery? How will it be recycled or repurposed? As you connect leads, route conductors, and arrange cells in the containment system, you carefully line up the wire runs and shield the connectors, not simply for safety reasons that might otherwise lead to lawsuits, but because you care about the health of the user (juridical and ethical concerns). And finally, the aesthetic experience of the user is more than a nicety for marketing. Images communicate and can engender an appropriate sense of trust in the user, guiding us toward both aesthetic (harmony) and pistic (faithfulness) normativity.

The book *Responsible Technology* refers to the design of a technological object that carefully takes into account all aspects of experienced reality as "sufficient design."[17] Sufficient design is an important starting point for engineers. It avoids the narrow blinders of designing only for efficiency in terms of energy, scale, cost, or any other single aspect of created experience. Sufficient design recognizes that creation is interconnected and that we should shape and form it holistically, but it can still fall short of developing technology that seeks the fullness of creation's potential. Holistic design that focuses on the artifact and its multiaspectual interconnections can help faithfully develop what is made, but this development can be strengthened by a parallel assessment of creation's dynamic trajectory. We do this by asking, what is it that creation and culture are to become? How, as a result of this technology, does the creation become fuller?

If the goal of designing for wholeness is sufficient design, then the goal of designing for fullness is *attentive design*. Attentive design constantly remembers where God has led us in the past (history) and stays alert to where God (by his Spirit) is leading us in the present and future (eschatology). Whereas sufficient design seeks to remove the reductionistic blinders associated with a given time and place and culture, attentive design steps back out of time and place and culture. Sufficient design emphasizes God's unchanging norms for his creation, while attentive design emphasizes God's ongoing dynamic relationship with his creation. Design for fullness holds loosely to our contemporary conclusions of responsible technology, recognizing the need for continual reassessment of God's will for technological development as each new generation fixes their eyes on Jesus, the architect of a renewed community en route to the new Jerusalem. Creation was created to change as God's creatures continued to live in his light and follow his leading.

We know from history that even though battery system design was the central technical challenge in every era of electric vehicle development, sufficient design alone was not enough to guarantee success. Designers today, working to achieve both wholeness and fullness, need a historical perspective. Even in the abbreviated story told above, we see that achieving the goals of sufficient design requires us not only to examine the entire system of technical artifacts (batteries, wheels, suspension, etc.) necessary to build an electric vehicle, it also means we need to recognize that achieving automobility is presupposed to be the primary measure of "progress."

Looking back, we can get a better sense of what biblically guided, culture-shaping decision makers should have done. Because recreation was assumed as the leading purpose or role for the development of electric vehicles in the early years during which automobility was being defined, designers ought to have thought about what constitutes good play (aesthetic life) and asked questions such as, Should technology encourage individual play over communal play? Should recreational technology be developed that excludes members of certain social and economic classes? What level of material resource consumption is justified for good play? How are users enriched physically, biotically, socially, and psychically by participation in a particular form of recreation? All of these questions need the illumination of Scripture to be answered.

Scripture does not prescribe play, but it does provide stories about what God desires of his people in community. Communities of shalom are characterized by a diversity of people from every nation, by generosity and shared resources, and by enjoying both work and celebration. There are likely aspects of this automobility ideal that resonate with a biblical notion of play—its desire to take delight in wilderness and parks in which all creatures can be celebrated, the creative playfulness in the artisan who crafts an engine or electric motor in a new way that eliminates costly maintenance, the joy that the automotive engineer finds in this work, and the possibility of outings that bring families from different regions for joyful reunions and feasts.

But we may also see in automobility an ideal of recreation that favors the wealthy and powerful, and excludes those who are neither. We might see in automobility an ideal that ignores the value of health and physical exercise for the lure of surpassing speed boundaries and diminishing space limitations. In other words, before we can address what a normative design might have been for early electric vehicles and their battery systems, the question of what is "good play," in the light of Scripture, for a particular place and time calls out for an answer. As automobile technology continues to develop (culturally unfold), the answers given to this question should move beyond the early recreational (aesthetic) reasons for its development and ask how juridical, ethical, social, economic, kinematic, and spatial questions will shape our automobility choices as they relate to both transportation and recreation.

Retelling our stories can help us understand how to seek fullness as we develop technology and achieve wholeness in our designs. Looking at the history of electric battery maintenance, for example, leads us to assess the process more broadly. Does the maintenance of the new technology compare favorably to the maintenance of current alternatives and past options? How was the work of those who maintained the horse and carriage infrastructure impacted? How can we preserve appropriate continuity while introducing change?

This raises another example. Knowing the story of how the electric vehicle affects infrastructure pushes us to ask what role the government should play in the development of public services. Should the government maintain, subsidize, and regulate roads, but not trolley systems, stables, or railroads? Who decides where roads begin and end? Who is responsible for public infrastructure, and where are the boundaries of municipal, state, and federal development? Do highway taxes ensure or infringe on our seeking of justice? Automobility assumes answers to these questions, even when there has been no conscious consensus. Congestion caused by trolley, horse, pedestrian, and automobile cannot be solved by holistically designing a battery or by merely conducting an analysis of traffic patterns and routes. Automobility also presupposes beliefs about gender, age, and social status. Who the intended users are and why they would use a technology should be factored into the design. The experience and awareness of pollution in the nineteenth century is not the same as the experience and awareness of environmental threats in the twenty-first century. History helps us recognize that normative stewardship in a particular era must be continually reassessed before we begin designing a component or system. All of this takes the establishment of a cooperative culture of transparent communication, something that did not happen among the early electric vehicle companies. Without it, the standardization or system integration necessary to sustain electric automobility was impossible. In fact, the stories of infighting reveal not only a lack of transparent communication, but a lack of concern for the corporate neighbor and an overall mistrust of individual or institutional neighbors.

In summary, it is probably evident at this point that these meta-technical questions—which ultimately determined the early fate of electric vehicle technology, for better or worse—were unlikely to have been part of the design decisions made for early electric batteries. However, just because they were not a conscious part of the process, does not mean they were not presupposed in the design process.

History will never answer the question of what responsible technology is and what it ought to be, but it does frame the questions needed for us to discern what wholeness and fullness should look like. This twofold biblical metric (wholeness and fullness) for progress needs to be revisited in each new generation. Every community needs to look to the past (where we have been) in the light of the future (where God is leading us) to make technological choices that faithfully respond to the Creator. Responsible historical decisions related to technology may be different from our contemporary decisions. The same technology may be rightly judged progressive in one era and regressive in another.

Design for fullness and wholeness is iterative. It is never complete. Returning to our electric vehicle case study, once we have worked toward sufficient design for the battery system, we must initiate attentive design and do the same for infrastructure and transportation planning questions, followed by recreation questions, gender and social roles questions, environmental questions, communication questions, and public policy questions. All questions are addressed in the past, present, and future narrative. In doing so, we will likely find overlapping, parallel, and intersecting things to consider. Only when we have the integrated whole in view can we, guided by the Holy Spirit, begin to work toward development that points to the fullness of renewed living as it is reflected in Scripture. Arriving at this vantage point of integration and future consummation, there is no list or formula to follow; we must simply allow the fullness of creation to be viewed through the lens of the kingdom's fullness yet to come. Scripture assumes that we have cultivated the imaginations needed to do this as we simultaneously immerse ourselves in the world (historically, aesthetically, analytically, lingually, etc.) and are guided by Scripture and nourished by God's Word. Engaging Scripture—like technological culture making—is a practice, not a procedure.

History, then, helps us more clearly see the religious beliefs of progress we have inherited. History humbles us. History shows us not just what happened but what we value or devalue. It reveals the direction of our hearts. Biblically defined progress will always be measured by the orientation of the minds and hearts of engineers, consumers, and drivers of any technology.

The orientation of our hearts and our assumptions of progress give rise to our cultural vision of the future. The next chapter examines how our historically conditioned imaginations give rise to both complementary and conflicting ideals for the future of technology.

CHAPTER EIGHT

TECHNOLOGY AND THE FUTURE

It's tough to make predictions, especially about the future.

YOGI BERRA

A field guide about technology can provide a richer perspective by zooming out beyond our immediate locale. The previous chapter looked back into history, and now Derek Schuurman helps us peer forward to imagine the future of technology. By pointing our design efforts to our broader eschatological hope we can avoid undue utopianism (triumphalism) or pessimism (hopelessness).

As a young boy, I recall leafing through old issues of *Popular Mechanics*, a science and technology magazine targeted for a general audience. As an aspiring engineer and electronics hobbyist, I found the illustrations on the magazine covers from the 1950s and 1960s compelling, portraying a future of flying cars, jet packs, and personal helicopters, and offering a variety of ambitious do-it-yourself projects.[1] In chapter one of this book we discussed how dreams inspire technology and engineering. Browsing the magazine covers of *Popular Mechanics* in this era illustrates the optimistic dreams of a future full of fantastic machines and technical possibilities.

The 1960s brought spectacular scientific and technological breakthroughs in atomic energy, space travel, and medicine. These developments fueled optimism in the prospects of scientific and technological change. This optimism was reflected in voices like physicist Alvin Weinberg, who coined the phrase "technological fix," a term which suggested that technological solutions provided the best way to address social, political, and cultural problems.[2]

But these technological advancements coincided with a time of social upheaval, protest, and Cold War fear. The early 1960s saw the consequences of thalidomide, a drug given to mothers for morning sickness that later led to thousands of tragic birth defects. Rachel Carson's influential book *Silent Spring* pointed to the environmental consequences of pesticides, helping drive a growing antitechnology movement.[3] She was lauded for raising awareness about "the calamitous possibility that our shortsighted technological conquests might destroy the very source of our being."[4] Rather than looking for a "technological fix" to our current problems, a growing number of people began to see technology as the *source* of even greater problems.

Both optimism and pessimism toward technology have shaped science fiction stories in movies and on television. In this chapter we will explore the hopes and fears toward technology as they have been expressed in popular culture and contrast them with a biblical view of technology and the future.

TECHNOLOGICAL OPTIMISM

Technological optimism was on full display at the 1964 World's Fair in New York City which featured exhibits like "Futurama," a narrated ride sponsored by General Motors about the not-too-distant future. The ride portrayed moon explorations, Antarctic weather stations, farming on the ocean floor, and futuristic visions of the cities of tomorrow. The fair's "Carousel of Progress" exhibit was later moved to Walt Disney World in Florida. It featured a rotating animatronic show depicting technological changes experienced by a family through the twentieth century. Each time the carousel revolved, an optimistic theme song was played with the words, "There's a great, big, beautiful tomorrow, shining at the end of every day. There's a great big, beautiful tomorrow, and tomorrow is just a dream away!"[5] In chapter one, we discussed how dreams inspire technological innovation; now, we need to discern the basis for dreams about creating a "great big, beautiful tomorrow."

It was also during the mid-1960s that the original *Star Trek* television series debuted. The show optimistically portrayed an advanced future through the adventures of a starship crew whose mission was "to explore strange new worlds, to seek out new life and new civilizations, to boldly go where no man has gone before." Each episode presented situations in which the characters combined technology with human wits to overcome threats and challenges they encountered along the way. Although the starship was equipped with advanced weapons, these were only used in self-defense. In all their interactions, the crew was guided by the "prime directive," a directive to not interfere or impose themselves on other civilizations. The show addressed contemporary social issues such as race, war, religion, and human rights in the setting of a future society that valued peace, prosperity, and knowledge.

Despite the commendable goals celebrated by the show, it was a future portrayed without God. The show reflected the worldview of the series creator, Gene Roddenberry, who "for years associated himself with humanism, a vague philosophical view that emphasizes a move away from God and toward the importance and achievements of the human being."[6] This humanistic view is clearly articulated in an episode where Captain Kirk and Spock find themselves transported back to New York City during the 1930s depression.[7] They encounter a street mission run by Edith Keeler, who later develops a romantic connection with Captain Kirk (such romantic entanglements with Kirk were a common plot element). At one point in the story, Keeler stands in front of a group of men gathered for a meal in the street mission and exhorts them with a rousing speech about the future:

> Now, I don't pretend to tell you how to find happiness and love, when every day is just a struggle to survive. But I do insist that you do survive, because the days and the years ahead are worth living for. One day soon, man is going to be able to harness

incredible energies, maybe even the atom, energies that could ultimately hurl us to other worlds in some sort of spaceship. And the men that reach out into space will be able to find ways to feed the hungry millions of the world and to cure their diseases. They will be able to find a way to give each man hope and a common future, and those are the days worth living for.[8]

These lines, spoken through a social worker in a soup kitchen in the 1930s, exhibit an unabashed technological optimism about the future, an optimism undergirded by a trust in technology to solve human ills. This quote succinctly captures the ideological underpinnings of the original *Star Trek* series, one that appealed to a generation of engineers. As mentioned in chapter one, a model of the starship *Enterprise* was placed in the Smithsonian National Air and Space Museum as a symbol of the dreams it inspired. Whether or not engineers happen to be "Trekkies," it is not uncommon for engineers to have an optimistic posture toward technology and the future. Perhaps this is due in part to self-selection—those who are pessimistic about technology are not likely to become engineers.

Such optimism persists to this day. The book *Infinite Progress* exemplifies this optimistic view in its subtitle: "How the Internet and Technology Will End Ignorance, Disease, Poverty, Hunger, and War."[9] While hope for restoration is one that Christians share, it's important for us to discern the true source of such hope. This subtitle has religious undertones, pointing to technology as the source for future hope. Such a trust in technology finds its roots in self-sufficiency and human autonomy, like a modern-day tower of Babel as described in chapter six. In contrast, Christians have a hope for the future that is not rooted in what humans can do but rather in the finished work of Jesus Christ. We look forward to a new heaven and earth in which "he will wipe every tear from their eyes. There will be no more death or mourning or crying or pain, for the old order of things has passed away" (Revelation 21:4). Ultimately, the holy city is one whose "architect and builder is God"

(Hebrews 11:10). Our technical work, then, should flow out of this hope, moving the focus from advancing technology for its own sake to directing technology in obedience to God and in service to our neighbor and the rest of creation.

Transhumanism. When I was growing up, there was a popular television series titled *The Six Million Dollar Man*.[10] The show told the story of Steve Austin, an astronaut who was severely injured in a catastrophic accident that caused him to lose both legs, an arm, and an eye. The series begins with Austin being equipped with "bionic" prosthetics that gave him superhuman capabilities. I remember the Steve Austin action figures which became very popular in the 1970s and included features like a "bionic grip" and a "bionic eye" you could peer through from the back of the head. Each television show opened with the memorable lines, "Gentlemen, we can rebuild him. We have the technology. We have the capability to make the world's first bionic man. . . . Better than he was before. Better, stronger, faster." This last sentence captures the technological optimism of a philosophical movement called *transhumanism*.[11]

Simply stated, transhumanism is a worldview that seeks a technological fix to current human limitations. Its goal is to enhance human intelligence and physical capabilities beyond natural human limitations through science and technology.[12] Transhumanists are technological optimists who pursue this goal through "bio-hacks" that include genetic enhancements, memory augmentation, and anti-aging techniques. Most people agree that technology shapes us, but for transhumanists such shaping is the goal rather than the unintended side effect.

It is not hard to find transhumanist themes in popular culture. The backstories of many popular superheroes describe how they gained special powers through technological or biological enhancements. For example, Iron Man and Batman gained special powers through technological augmentation; for the X-Men, Spider-Man, and the Hulk, special powers came through genetic mutations. Such stories of superhuman capabilities are not new and can be found already in Greek mythology.

The transhumanist movement includes many presuppositions about what it means to be human and what it means to flourish. A key value of transhumanism is the notion of "morphological freedom" which suggests that people have a right to modify themselves in any way that they see fit.[13] Many familiar technologies modify humans to *restore* certain capacities. For example, eyeglasses are a type of technological augmentation intended to restore vision. Likewise, pacemakers, dentures, artificial limbs, and hearing aids restore regular human functioning. But, as author Jacob Shatzer explains, advocates of morphological freedom have bigger changes in mind: "Not the ability to wear glasses or have surgery, but to have a tail if you want to."[14] Authors like Francis Fukuyama draw a distinction between "therapy" and "enhancement," even though the distinction is not always clear.[15] Creating technology to restore human capacities is a way to show love for our neighbor, but technical efforts aimed at changing humans for the purpose of enhancement are another matter. The Christian philosopher Craig Gay suggests that while we can "strive to improve our circumstances" we must "rule out seeking— impatiently and by means of technology—to redeem ourselves now."[16] He cautions that "just as impatience may well have aroused the original sin, so impatience must continue to drive us away from the divine purpose."[17]

The goals of transhumanism even go beyond enhancing human capacities; they seek to overcome death itself. Ray Kurzweil, an accomplished computer scientist and inventor, has suggested that someday we will be able to upload our brains into a computer and live forever, no longer limited by our mortal bodies. He believes that this will be possible once neuroscientists are able to map the human brain and develop sufficiently powerful computers. This aspiration has echoes of Gnosticism, an early church heresy that arose from both Greek and pagan influences that viewed matter as evil and the spirit as good. In the words of Kurzweil, "We won't always need bodies, in a virtual world virtual bodies will be good enough."[18]

Such a view assumes a profoundly mistaken notion of what it means to be human. It reduces being human to being able to simulate the interaction of particles in the human brain in a computer. In a sense, the very first temptation in the garden was the suggestion that humans could become like God. In his book *Playing God*, Andy Crouch observes that "every idol makes two simple and extravagant promises. 'You shall not surely die.' 'You shall be like God.'"[19] A biblical view of humanity is one that recognizes our creatureliness, acknowledges that we are not autonomous, and that our "help comes from the LORD, the Maker of heaven and earth" (Psalm 121:2). Any attempt to build a life apart from God is futile. Thankfully God came to us in the person of Jesus Christ, the one who has already conquered sin and death.

Christ's incarnation shows us the value God places on our physicality and humanity. We should not be so eager to shed our current humanity. In fact, Jesus did not shed his body after his work on earth was done but ascended into heaven in human form. One day, we too will have new bodies, not through transhumanist technology but through Christ, "the Word who became flesh" (1 John 3:2). In the words of Craig Gay, "the incarnation seriously puts an entirely new spin on what we ought to use our technologies for."[20] He suggests that we should pursue technologies that enhance our actual experience of ordinary reality by giving us time to really listen, think, reflect, and "dwell richly in those places where we are most at home."[21]

TECHNOLOGICAL PESSIMISM

While the original *Star Trek* series was largely optimistic, more recent science fiction narratives portray a far bleaker picture of the future. The opening scene of the movie *Terminator* presents a battle scene between humans and an army of killer robots. The camera focuses on a pile of human bones and skulls as they are crushed beneath the tank treads of an advancing robot. The camera then pans to a battle scene of robot warriors firing bursts of laser cannons at a small group of humans desperately scrambling for cover like insects. The robot army is under the control of Skynet, a computer program that was intended to defend the United States against attacks but instead went rogue, unleashing nuclear war and turning on its human masters. In this brief scene we witness a stark world of technology run amok. Another film, *The Matrix*, depicts humans unknowingly kept captive in a virtual world while their physical bodies are stored in pods to provide power for the machines. The television series *Battlestar Galactica* tells the story of a ragtag remnant of humanity battling robots called Cylons who are attempting to exterminate the human race. Each of these tales is derived from some variation of the "Frankenstein narrative" in which technology turns on its human creators to destroy or enslave them.

Science fiction has provided an effective way for storytellers to dream about the consequences of our ongoing technological developments, for better or for worse. Why are these stories important? Steve Garber writes, "We have to pay attention to the novelists, filmmakers and musicians who are culturally upstream, as it is in their stories, movies and songs where we feel the yearnings of what human life is and ought to be. . . . They are fingers to the wind. Why? Artists get there first."[22]

> **TECHNOLOGICAL DETERMINISM** is the belief that technology is an autonomous force beyond our control.

Many contemporary science fiction stories imagine a grim future and reflect a state of hopelessness. Some of these narratives are informed by a sense of *technological determinism*—the notion that technology is an autonomous force beyond our control. This view is reflected in the phrase "Resistance is futile" that the alien Borg broadcast to their victims in *Star Trek: The Next Generation*. The Borg are a collective of cybernetic organisms that try to assimilate everything in their path. Like the onslaught of the Borg, technological determinism por-trays technology as an unstoppable force.

Science fiction as a genre can be instructive for Christians because it shows what misdirected technology and its consequences can look like. These stories often capture our attention more effectively than philosophical arguments. Thankfully, for most engineers, the conse-quences of poor engineering may not be as planetary in scope or as dramatic as those depicted in fiction. Even so, fiction can help convict engineers of the need to be sensitive to the social, environmental, eco-nomic, and justice consequences of their design decisions.

It's important to note here, however, that many dystopian science fiction stories portray technology as the root of society's problems. This is sometimes referred to as the "progress trap"—the notion that human ingenuity inevitably leads to problems that we will be unable (or unwilling) to solve.[23] Author and civil engineer Samuel Florman refers to people who use technology as a scapegoat for our problems as "antitechnologists." He suggests that "the true source of our problems is nothing other than the irrepressible human will."[24] As Christians we cannot blame technology or anything else in creation for our predicaments; the problem lies in the human heart. In the words of theologian Al Wolters, "The great danger is always to single out some aspect or phenomenon of God's good creation and identify it, rather than the alien intrusion of human apostasy, as the villain in the drama of human life."[25]

Both antitechnologists and technological optimists are similar in that they elevate one aspect of creation, namely the latent potential for technology. While technological optimists look to technology as the solution to all our problems, antitechnologists look to technology as the source of all our problems. The truth is that both our salvation and the root of our problems are to be found elsewhere. Wolters writes, "The Bible is unique in its uncompromising rejection of all attempts . . . to identify part of creation as either the villain or the savior."[26]

Figure 8.1. Image from the movie *Wall-E*

To be fair, some science fiction stories are more nuanced, exploring the possibility of hope if human beings are willing to exercise freedom and responsibility. The movie *Wall-E* is an encouraging tale about hope in the presence of environmental degradation and dehumanizing automation. The movie opens in a dystopian future where humans have polluted the earth (see fig. 8.1) and are relegated to a large ark-like ship. Aboard the ship are obese, passive humans who are pampered by automated systems that control every aspect of their lives. In a climactic scene, the captain, who is unaccustomed to walking, waddles over to the control panel and wrestles control away from the ship's automated system in order to steer a course for earth. The movie ends as humanity returns to earth with a fledgling young plant, hinting that the proper exercise of our God-given responsibility can help lead to flourishing for humanity and the planet.

A BIBLICAL VIEW OF THE FUTURE

The biblical story begins in a garden but ends in a city. Amazingly, the Bible suggests that some of our cultural artifacts will also appear in the new heavens and earth. Revelation 21:24-26 speaks about how the "glory and honor of the nations" will be brought into the new Jerusalem.

Likewise, in Isaiah 60 we read of the "riches of the nations" being brought into the city of Zion. These include raw materials, like precious metals and lumber, but also cultural artifacts like the ships of Tarshish. Richard Mouw suggests that these ships were "instruments of pagan commercial power," but that they somehow appear in Zion repurposed for service to the Lord.[27] In Micah 4:3, we read that swords will be beaten into plowshares and spears into pruning hooks. This is an image of technology, originally directed toward war and harm, being redirected toward peaceful purposes. Indeed, plowshares and pruning hooks are tools that remind us of the cultivation and stewardship to which we were originally called in Genesis 1. These passages paint a biblical picture of the future, one that includes technology, and one that Christians can eagerly anticipate.

In the meantime, Christian engineers are called "to make some imperfect models of the perfect world to come."[28] As described in chapter two, the unfolding of technology is a response to God and the original cultural mandate. Technology is not an autonomous force beyond control, but rather a cultural area in which we are called to exercise freedom and responsibility. Although sin has distorted our technical dreams and activities, the call to respond obediently persists. The biblical view of the new heavens and earth, complete with redeemed technology and other cultural artifacts, underscores the legitimate place of technology and design. With God's help, the faithful engineering work done in cubicles today can be signposts of God's coming kingdom— forging technologies in response to God's call to care for the earth and its creatures and which demonstrate love for our neighbors.

CONCLUSION

We have explored how technology is often perceived as the controlling factor that will determine the future, for better or for worse. Optimists see technology as ushering in a utopian world, with transhumanists seeing technology as a means to unlimited human potential. On the other hand, technological pessimists and antitechnologists blame technology for many of the problems that plague humanity. Technological determinists go even further, suggesting technology is its own autonomous force that will inevitably control our destiny.

Our perception of the future has a profound impact on how we live today, including how we approach technology and design. Science fiction stories can help us recognize the dual pitfalls of undue technological optimism on one hand and the fatalistic fear of technology on the other hand. Viewing technology as the primary pathway to a better world (or to a better human) will drive us to grasp for technological solutions that will ultimately fall short, like any other idol. However, a pessimistic view of technology misdiagnoses the problem, one that begins with the human heart. Technology is a gift that ought to be gratefully unfolded and exercised with responsibility, not made a scapegoat for our problems. The Bible depicts a future not centered on our technological progress, but one in which God himself will usher in the new heavens and earth, one which "will be filled with the knowledge of the glory of the LORD as the waters cover the sea" (Habakkuk 2:14).

The field of engineering is an important calling for those who want to develop technologies that help human beings and the entire planet flourish and bring glory to God. As the movie *Wall-E* suggests, rather than passively allowing technology to take its course, we must respond by pursuing responsible technology. God has given us "the ministry of reconciliation" (2 Corinthians 5:18) and this extends to all aspects of life, including engineering and technology.

The previous chapter took a historical look back whereas this chapter focused on looking forward. Regardless of whether we look forward or backward, we can see how theological and philosophical perspectives inform our view. The last two chapters now take a more personal and practical view of how one can be faithful and honor God in our technical work.

CHAPTER NINE

MUST WE LEAVE OUR NEURAL NETS TO FOLLOW HIM?

I see nothing inconsistent in monks of Christ building a flying machine, although it would be more like them to build a praying machine.[1]
WALTER M. MILLER JR.

Many people working in technology are pursuing a professional career, such as an engineer, scientist, or manager. As Christians, we might also feel a vocational tug toward the mission field. In this next chapter, Steve VanderLeest makes the case that these are not mutually exclusive desires, hopefully demonstrating that technological work is a legitimate Christian calling.

Emily left engineering to be a missionary. She had started college as an engineering major after her high school teachers and vocational counselors encouraged her to pursue a technical career. It was sound advice based on her interests and skills. She excelled in math and science classes. She had a creative flair with technology and liked to solve practical problems. She was an avid member of a local robotics team that had placed well in a number of competitions. Engineering was a natural fit. Emily worked hard, kept a high GPA, and still had a little time for fun with friends. She was satisfied that she had a bright career path ahead.

However, a gnawing doubt soon began distracting her from that promising future. Her pastor and the youth leader at her church frequently challenged her to think about her gospel witness, about a higher calling to tell people about Jesus. Although they never directly condemned working in the "secular" world, it was clear that they believed that the more noble calling was to be a pastor or missionary. Because Emily wanted to put God first in her life, she started to have concerns about choosing a well-paying career that might feed a greedy, self-serving lifestyle. She worried that she would be selling out to a corporate world that cared nothing about real people. She decided to step out in faith and give up her dream career in order to follow Jesus.

While the previous chapters have taken broad theological, philosophical, or chronological perspectives, the purpose of this chapter is more personal: to demonstrate that professional work in technology is a legitimate Christian calling. Emily was mistaken. It is not only missionaries who do God's work. She did not realize that she could also serve God and witness to the gospel as an engineer.

Emily was not seeing the whole picture. God, who is sovereign over all, "rule[s] over all the kingdoms of the nations" (2 Chronicles 20:6). All legitimate professions can serve him and, in fact, only engineers can witness to the gospel in certain areas. Engineers can carry out the commission of Matthew 28:18-20 by connecting to people with whom they work and by loving their neighbors through the development of new technologies that improve health, save lives, provide security, enable creative expression, and raise living conditions.

At the start of his ministry, Jesus calls the fishermen brothers Peter and Andrew to follow him: "At once they left their nets and followed him" (Matthew 4:20). However, we should not infer from this that everyone must drop their vocation in order to follow Jesus. It might seem that the modern version of this story would require that programmers who use neural nets put aside their software to follow Jesus. Yet, reading further, though they leave their nets to follow Christ when they become disciples, we also find Peter and Andrew fishing in later chapters. Likewise, those in technical disciplines who are called to be disciples of Christ must put first what Christ put first (i.e., God's love for the entire cosmos, John 3:16). That doesn't mean, however, that we can't work within our areas of technical expertise, that we shouldn't continue to "fish" with our neural networks. In fact, we serve Christ concretely through our work and within our professional interactions. In this chapter, I first examine why some Christians mistakenly scorn technology and then look at how we can serve God through technology.

SCORNING TECHNOLOGY

"Christians should avoid music." That's an absurd statement, right? There might be some types of music that must be treated with care and discernment, and there might be songs that are best avoided by most Christians because of their trashy lyrics. Nevertheless, to avoid all music because of some bad pieces would be like avoiding all fruit because of a few bad apples. While this example may seem a bit of a stretch, it is no different from calling for believers to avoid all technology. Several reasons may drive some Christians to make such a judgment, causing faithful and devout believers to think they should avoid technology altogether. It could be a misguided attempt to return to Eden, a distortion of the notion of holiness, or a mistaken correlation of technology with its possible sinful effects. Let's look at each of these reasons in turn.

Misguided return to Eden. Aversion to technology might be a desire to return to a time reminiscent of the Garden of Eden. Such a view mistakenly sees any human intervention in nature as evil, a scar on a pure landscape. Some well-meaning biologists, ecologists, and nature lovers have made the claim that technology ruins the earth. They see signs of corruption in ecological disasters such as large oil spills and in the visible scars on the land left by strip mining. Frankly, in cases like these, they are right. Technology is a powerful tool. Used improperly, it can drastically damage our environment, even on a global scale. However, even multiple examples of abuse do not necessarily mean that technology itself is inherently evil. We can find many examples of beneficial technology that offer counterexamples to this way of thinking.

Wistful dreams of returning to Eden are prevalent in our world. Many people prefer natural products, reflecting an affinity toward the goodness of creation. This preference for a pure environment may open us to the danger of making nature an idol. Some Christians who care deeply about God's world point to the pure, natural state of the original Garden of Eden as the standard by which to judge the impact that the modern world, and especially technology, has on the natural environment. That impact is judged to be largely negative.

If the job of perfect humans in a perfect garden world was to keep and tend, then highways, ships, and oil-burning machines all seem to be cruel displacements of that garden. So, let's think together about how an affinity for the good creation makes people averse to the artificial and altered materials that we associate with technology. God's good creation is meant to be protected and meant to be cultivated. The goodness of creation is obvious from the biblical story, and Eden is an idyllic picture of that goodness. However, one cannot stop reading the biblical narrative after the first two chapters of Genesis. Genesis begins the story of the unfolding of the gift of creation, and that unfolding continues throughout history. That unfolding also includes us, as God's stewards of his creation. And that unfolding includes technology—it is not an evil that opposes the good of creation, but rather is part of that creation.

We see evidence of technology's place in God's plans for his creation in the picture the Bible paints of cultural development and the cultivation that occurs as history moves from the original garden in early Genesis to the city of the new Jerusalem envisioned in Revelation. We live in between those times, giving us the opportunity to consider how technology can aid rather than harm our environment. One example is the use of solar or wind as the source of energy for heating and cooking rather than burning wood, a traditional method that heavily pollutes the air.

Misguided distortion of holiness. Aversion to technology might also come from a distorted notion of holiness, the virtue of keeping oneself separate from worldly evils. Christians are called to be holy, set apart for God. The Amish offer an example of people eschewing technology for religious purposes. They restrict smartphones and do not have televisions in their homes. However, the Amish do not avoid all technology. Rather, they are thoughtful and methodical about what technology they choose to incorporate into their communities. Consider that the stereotype of Amish culture is the horse-drawn black buggy riding slowly along the side of the road, often with an orange triangular warning sign on the back as a caution to fast-passing automobiles. Although dated, buggies are nonetheless technology. The harness on the horse, the steel wheels of the carriage, and the mechanical braking mechanism are fine examples of technology that the Amish have intentionally adopted. Selectivity, rather than complete avoidance of technology, more accurately describes their way of life.

Amish selectivity with technology may be a lesson for all of us—holiness demands deep discernment if we are to recognize the ill effects that might arise through the misuse of technology. We might not be as selective as the Amish in forgoing most technology, yet we could aspire to be more thoughtful in the choices we make and more thoughtful with how we use the technology we choose to design and to own. Designers must be especially discerning, as noted earlier in this book, considering carefully what ill effects might arise through designs that enable or encourage misuse. Perhaps we could learn something about the intentional engagement of technology from the Amish—not by avoiding it, but by more discerning use and development of it.

Being holy means setting ourselves apart and turning our hearts toward God instead of pursuing sinful desires. The apostle John says, "Do not love the world or anything in the world. If anyone loves the world, love for the Father is not in them. For everything in the world—the lust of the flesh, the lust of the eyes, and the pride of life—comes not from the Father but from the world" (1 John 2:15-16). Discerning use of technology must turn us from sin and toward God. That means we must evaluate particular technologies based on their merits, making sure that we are not loving any particular technological convenience above God or before our neighbor—a love that would turn that device into an idol.

One way to detect technological idolatry is to consider how much time and money we spend on our personal gadgets, and furthermore, to consider whether we use those personal gadgets for mostly selfish reasons or for the building up of others. This is the same kind of discernment that we must employ toward the world and the culture around us—not only in technology but also in music, literature, law, food, or movies. The world belongs to God. We are called to claim technology, as everything else, for God as we watch for the day of the Lord described by the prophet: "On that day HOLY TO THE LORD will be inscribed on the bells of the horses, and the cooking pots in the LORD's house will be like the sacred bowls in front of the altar. Every pot in Jerusalem and Judah will be holy to the LORD Almighty" (Zechariah 14:20-21).

Misguided correlation with sin. Aversion to technology might also result from mistakenly correlating technology with its sinful effects. Technology is not inherently evil, but it certainly can be warped into an idol. King Uzziah provides us with a case study. We read in 2 Chronicles 26 that Uzziah started his reign quite young, at the age of sixteen. At first "he did what was right in the eyes of the LORD" (v. 4). The biblical description of his reign points to a variety of advancements in architectural technology, such as fortified towers (v. 9) and the agricultural technology of cisterns to water livestock (v. 10). He even invented certain types of military technology: devices to shoot arrows or hurl stones from towers (v. 15) for the defense of Jerusalem. All of these examples appear while Uzziah was right with God. But then Uzziah turns away from God. Scripture says, "After Uzziah became powerful, his pride led to his downfall" (v. 16). It's important to note here that it is not the technology itself that leads him astray but rather the misdirection of the power that technology provides. Pride is a vice that Christians are called to abhor, and in this passage, they are called to abhor the sin, not the technology that facilitated his pride. Misdirected power derived from technological devices corrupts and can lead to injustice. We are called to abhor injustice, but rather than also scorning the organizations or technology that may have contributed to the sin of injustice, we are called to redeem them, turning them away from evil and toward the good.

It may be important to say here that although technology is not intrinsically sinful, it can powerfully enable and amplify sin. Technology comes with responsibility and, as discussed in chapter three, the power it offers can create tendencies and preferences toward certain uses, sometimes unintentionally. This puts a special responsibility on technology developers to be holy and discerning as they design devices that help users to flourish. Scripture does not suggest that we avoid technology. Its message is more nuanced as we'll see in the next section, which briefly surveys examples of technology in Scripture.

BIBLICAL EXAMPLES OF TECHNOLOGY

The Bible does not teach us that technology is evil. On the contrary, it paints a picture of cultural development and cultivation of creation moving from the original garden in early Genesis to the city of the new Jerusalem envisioned in Revelation. Throughout the biblical story, we find many examples of technology that the Bible treats positively. Shortly after the creation story, the two sons of Lamech make early technological devices. Jubal was "the father of all who play the harp and flute" (Genesis 4:21), musical instruments that required fine craftsmanship. Tubal-Cain, "forged all kinds of tools out of bronze and iron" (Genesis 4:22). Bronze is an alloy made from copper and tin, showing a certain sophistication in technological metallurgy. Noah designed and built the ark (Genesis 6), a complex structure, following direct instructions by God—a boat that was likely larger than any other at the time and a technological accomplishment that must have required a large construction zone and was likely a sight to see! Abraham and Isaac engaged in civil engineering projects when they dug wells in Genesis 21:30 and Genesis 26:18-33.

Bezalel was filled with the Spirit of God to build the tabernacle "with wisdom, with understanding, with knowledge and with all kinds of skills—to make artistic designs for work in gold, silver and bronze, to cut and set stones, to work in wood, and to engage in all kinds of crafts" (Exodus 31:3-5). King Solomon architects and builds the temple in 1 Kings 6, and Huram skillfully does the bronze work required for many of the temple furnishings. King Uzziah employs "skilled workers" in 2 Chronicles 26:15 (NRSV) to design military machines to defend the city. Nehemiah rebuilds the wall around Jerusalem after the return from exile. By and large, these examples give positive representations of technology.

Technology is portrayed positively in the New Testament as well. Jesus uses technology as the "visual aid" in many of his parables:

- ▸ Houses in the story of the wise and foolish builders in Matthew 7:24-27

- ▸ City gates in Matthew 7:13-14

- ▸ Cities and lamps in Matthew 5:14-16

- ▸ Fishing nets in Matthew 13:47-50

- ▸ Oil lamps in Matthew 25:1-13

- ▸ Lost coins in Luke 15:8-10

Although we may not often think of these as examples of technology, they really are the technologies of the time. Houses and cities require architecture and civil engineering; coins require technical know-how in metallurgy and manufacturing capability. The Bible also notes that Jesus was a carpenter's son, and Jesus himself was a carpenter. This tradecraft was a technology of that day. The elements that Jesus offers the disciples in the Last Supper are technology as well: the bread is processed wheat, and the wine is processed grape juice. Paul uses technological allegories such as the one in Ephesians 6:10-17 where he speaks of the armor of God.

These biblical examples describing God's people making their way in the world demonstrate that technology is a legitimate cultural activity and an integral part of what it means to be human. These stories also illustrate that technology is richly laden with embedded values and judgments. Christians must be discerning about the way we design and use technology, seeking to be redemptive agents in the technological development of God's good creation.

If technology is not inherently sinful and if we are not called to avoid it, how should we think about it? How do we design technology with Christian discernment? We do not need to leave our current career path in order to be faithful. The second half of this chapter gives some direction for those engaged in technological development. In the following sections, we'll use the lens of four significant biblical commandments to explore serving God in our design and use of technology: the Great Commission given in Matthew 28:18-20, the two great commandments given in Matthew 22:37-40, and the cultural mandate given in Genesis 1:28.

SHARING THE GOSPEL VIA TECHNOLOGY

In Matthew 28:18-20, Jesus calls his disciples (and us) to make disciples of all nations. In this section, we explore three ways Christians working in technology can make disciples as witnesses for Christ: witness in careers in general, witness in careers specific to technology, and witness through technology itself. After discussing making disciples as witnesses, we turn to making disciples through our acts of service to God and to our neighbor.

Vocational witness. In response to the Great Commission, we send missionaries or become missionaries ourselves to the thousands of unreached people groups around the globe that have not heard the gospel. It is no small task to make disciples of all nations. However, unbelievers are also close to home.

How close to home? Theologian R. C. Sproul pointed out that we witness "to the reality of the kingship of Christ in our jobs, our families, our schools, and even our checkbooks, because God in Christ is King over every one of these spheres of life. The only way the kingdom of

God is going to be manifest in this world before Christ comes is if we manifest it by the way we live as citizens of heaven and subjects of the King."[2] Opportunity to witness is all around us in our daily lives.

Our work is part of that daily living and a large group of unbelievers can be found within technical professions. They may or may not have heard the gospel, but if they do not believe, we are called to bring the gospel. Your job or profession allows you to get close to people in order to share transforming good news. Working together professionally offers opportunities for striking up conversations or friendships that make this possible. "Going to the ends of the earth," in this case, means reaching unbelievers cloistered within corporate walls that cannot easily be breached by a traditional missionary, spaces that can only be penetrated by a coworker and fellow professional. Some Christians might even be willing to take jobs they might otherwise not choose in order to gain proximity to those needing to hear the gospel.

Witnessing on the job requires wisdom and boldness as we proclaim Christ. Religious freedom laws in the US and many other countries protect the expression of religion, including at work, but these same laws also protect others from religious harassment. Sharing one's faith can be perceived as aggressive proselytizing and result in discipline or even termination. The legal limit to witnessing (technically termed "proselytizing") kicks in when it interferes with work performance of yourself or those around you.[3] Edna Ng lost her job at an engineering company, not because she was a Christian but because her proselytizing in the workplace detracted from her work performance. She also used company property for nonwork purposes (holding events to convert her coworkers) and was judged to have negatively impacted the performance of coworkers.[4] Edna may have believed that the Great Commission compelled her to witness in a certain way, but in her zeal, she didn't do the work for which she was being paid. She ultimately lost her job, lost her opportunity to witness, and perhaps gave coworkers a negative view of Christians. All of us employed in technical areas need discernment and wisdom as we find ways to be a continuing witness to the good news of Jesus Christ.

As we engage with coworkers, it's important to consider not only our witness through words but also through character and behavior. Actions speak louder than words. Let them know you are a Christian by your love—by your professional integrity, your compassionate behavior, and your wise design work. A preachy holier-than-thou tone creates antipathy, not openness. Christian professionals practice personal ethical behavior, speak and write with integrity and honesty, love and care for their fellow workers, treat company resources with respect and a sense of stewardship, and live joyfully in the face of difficult circumstances.

Such actions are certainly not limited to engineers, and much of this section applies to Christians working in any career. Whether you are an engineer, nurse, lawyer, plumber, or teacher, your character becomes obvious to your coworkers as they observe your day-to-day behavior on the job. In this sense, technical fields offer many opportunities for Christian witness. In the US alone there are millions of professionals employed in engineering, science, and technology fields. Sequestered inside corporate offices and technical labs, the gospel will rarely reach their ears unless a trusted colleague shares it with them. While Christians from a wide variety of vocational backgrounds can serve as missionaries in developing countries, only those with a highly technical education can serve as missionaries to this corporate mission field. Technical expertise opens doors, allowing us to be in the right place at the right time.

Following God's call in your technical profession includes more, however. Your witness within a technical vocation can go further in ways that are unique to your technical expertise.

Technical vocational witness. Our ability to work alongside others working in technical careers is enabled by our technical expertise. However, our technical expertise is not merely an enabler to witness that has nothing further to contribute. A closer look at career and employment builds on witness to bring a prophetic voice to the job. Christian philosopher Nicholas Wolterstorff points out that simply carrying out the tasks of a job is not sufficient. We must make those tasks worthwhile as well: "Remaining in that [occupational] role is not the thing which is to be done out of obedient gratitude; rather, the actions performed in that role are what is to be done out of obedient gratitude. . . . Each occupational role must either be made to serve the common good, or if in some case that cannot be done, then that role must be discarded."[5]

Technology-related careers, like many others, can raise distinct ethical questions, questions which chapters four and five noted should be considered from the much broader perspective of norms. As engineers, our personal integrity and care for our neighbor might make us question a design choice because it could lead to a design that does not honor God, for example, because it is unsafe or unjust. Our engineering expertise gives us the ability to recognize the impact of a particular choice. Our creativity, based on our Christian view of the world, helps us imagine alternatives that are safer.

An illustration from the workplace may help. I once worked for a demanding project manager, for whom I was testing a new single-board computer. He called me into his office one afternoon, just minutes before his weekly call with the customer who was eagerly awaiting this new product. "Eagerly" is a kind way of putting it. We were behind schedule, and the customer was angry and agitated. To make sure all was in order before the call began, my manager wanted a quick update from me about how my testing had progressed. I told him that the testing of the devices and interfaces on the front side of the board was coming along nicely, but another set of tests could not be run yet because we had not received a set of chips needed for the back of the board. The manager told me to report that everything checked out positively. My heart started racing. Was he asking me to lie? With only five minutes until the call, he nonchalantly repeated back my own words: the testing was "coming along nicely." Wasn't I confident that the backside device tests would pass once the delinquent chips arrived? Although I was reasonably confident in eventual success, many things could go wrong. I timidly indicated as much and suggested that if the customer asked about the testing, I could honestly report that all the tests I had run thus far were passing. I also said I could not lie to a direct question.

My manager seemed about to push further, but the phone rang and we began the call. After a few pleasantries, the customer started hammering my manager about the slipped schedule. My pulse quickened. They were keen to understand the details of the production run once the board had passed my testing and some environmental testing. For some reason, the customer was particularly concerned about this environmental testing and had many questions about it. Faster than seemed possible, the hour-long meeting was drawing to a close, and they had not touched on the subject I was dreading. Then it came. "Oh yes, what about the functional testing? Are you making progress?" This was the moment. I answered in the affirmative and then gulped, waiting for the follow-up questions that would put me in a tight position.

The harder queries never came. That was the only question for me and then the customer ended the meeting. I could hardly believe it. I was still on the hook, however. Those chips were already a couple of weeks late, and I was sure to be on the hot seat on the next week's call. Thankfully, the chips arrived the next day and, after populating our test board, I was able to quickly run the tests and report success during the next call. Here was a case where an ethical (normative) choice depended not only on discernment, but also on technical expertise to understand the implications, risks, and veracity of the conclusions I was asked to make.

Moral decisions are always intertwined with technical choices as engineers select design options that best take into account the good of the customer and the good of society. Creating a God-honoring design requires an understanding of the full, rich meaning of *good*; it requires an understanding of the implications of design choices. Making wise choices is one way that Christian engineers can witness to God's call to be faithful in our work, lending our voices as modern-day prophets to a technical world.

Being a prophetic voice can have unintended consequences though. An occupational hazard for Christian engineers trying to be obedient is that prophecy can appear as arrogance. We, like all Christians, are called to be holy but not "holier than thou." If our holiness turns into perceived or actual arrogance and pride, we damage our witness. Engineers, generally, are proud of what they do and are rather good at their jobs, making it hard for them to accept criticism. To earn credibility, you need to work hard and put others before self, demonstrating strong skills and a good work ethic. Only then is your witness to the gospel likely to be heard as sincere and authentic.

To sum up, if done thoughtfully and lovingly and based on biblical principles, our design and production of technological products can help us love our neighbor, promote justice, enable better stewardship, nurture healthy relationships, and more. If done without thought for broader principles such as design norms, our technology can unintentionally hurt neighbors, encourage injustice, waste resources, enable abusive relationships, and produce many other harms.

Witness in technology. A third way we witness builds on the two mentioned above: our access to technical work environments and our ability to speak prophetically to ethical choices. We can serve God by using technology to witness. Dan, a junior electrical engineering major, came to my office one spring semester, frustration on his face. He wanted to find an internship connected with a mission trip abroad. The only opportunities he could find were civil engineering projects, such as building water filtration systems or bridges. These projects wouldn't allow Dan to put his electrical engineering knowledge to work. The only internships he could find for electrical engineering majors seemed to be with big corporations, not mission organizations. I was quick to point out that big corporations are also a mission field for engineers, but I also gave him some ideas for using his skills with a traditional mission agency.

After looking a bit more, Dan found a summer internship with a Central American radio station, broadcasting the good news in Spanish. He helped improve their technology to produce a greater transmission range and he also added some new features to the simple computer system used as a user interface for the transmitter. Dan returned for his senior year invigorated by the impact he had made by using his engineering skills to reach more people for Christ. He had amplified, literally, the voice of one "calling in the wilderness" to reach those who had never heard the gospel before. Dan's work lived on for years after he left Central America, multiplying his witness many times. What a wonderful example of directing the power and amplifying effect of technology toward a kingdom purpose. Many technologies have opened a channel of communication for sharing the good news, from Gutenberg's press printing Bibles to websites providing Bible translations in hundreds of languages.

Thus far, we have focused on ways we can fulfill the Matthew 28 call to make disciples of all nations as Christian engineers, scientists, and technologists. Next, we will look at the ways our work in technology can also fulfill what Jesus calls the greatest commandments: to love our neighbor and to love God.

LOVING OUR NEIGHBOR TECHNOLOGICALLY

Technology can and should help us obey God's command to love our neighbor. It sometimes seems to do just the opposite. Technology has gotten a bad rap, sometimes deservedly so, by enabling bad behavior. From the early days of widespread television, parents have worried about whether allowing their children to watch too much TV might cause physical or intellectual laziness or put bad ideas in their heads. Later, video games became a big parental concern, especially violent games that might encourage violence in real life. However, humans have also been able to do much good in the world through technology. In fact, being thoughtful about technology can help us more richly fulfill our calling to love our neighbor. We encourage you to develop your own unique and creative ways to use technology to love your neighbor, prompted by some examples we provide here.

One way to love our neighbors who are less technically savvy is to point out to them the insidious uses of technology. Those of us who know technology know that it can dehumanize, defy biblical principles, or detract from God's glory. This does not mean rashly condemning all technology as evil but rather it means identifying how particular aspects of a technological system encourage specific human vices. By doing the hard work of a root cause analysis when our instruments do not serve us well, we love our neighbors.

We also love our neighbors technologically when we point out good uses of technology. Some gadgets and techniques enable humans to be truly human and, in doing so, glorify God. This means keeping technology in its proper place as part of God's creation and subject to the effects of sin.

Many technologies help us love our neighbor. An MRI machine does so by looking inside the body to identify the cause of a physical ailment. The usefulness of magnetic resonance to create images was first noticed in the 1970s, and then later came to be used for medical purposes, augmenting a physician's view of the patient's body beyond what the older technology of x-rays could provide.

A bicycle provides an inexpensive means of transportation that helps our neighbor get to work on time. Many Americans probably consider the car for this purpose, but loving our neighbor requires us to recognize a variety of financial situations as well as environmental impacts of the transportation technologies we provide.

A video conferencing app connects geographically dispersed family members through long-distance conversations that allow them to see each other. I was working on this chapter during the Covid-19 lockdowns when this became particularly important.

Wrinkle- and stain-resistant fabric enables our neighbor to look well-kempt throughout the day. Eyeglasses restore diminished sight. The list could go on and on—there are surely countless examples of loving our neighbor with technology. We love our neighbor technologically by designing new technology that addresses human problems in humane and God-glorifying ways. Virtually all people in today's world use technology in some way and can think about how they can love their neighbor with technology.

Those of us who are technology experts have a further calling, however. Not only can we put our instruments to good use but we also have the opportunity to design them well in the first place so that they are predisposed toward good uses. Furthermore, we can seek to influence codes and regulations to encourage and even require technological development that promotes justice, harmony, and all the design norms. You will likely come across such opportunities for influence, whether it is serving on a local drainage commission or serving on a national standards committee. We hope you jump at the chance to love your neighbor in this way.

LOVING GOD TECHNOLOGICALLY

Technology can and should help us obey the greatest commandment: to love God. In its first question, the Westminster Shorter Catechism asks, "What is the chief end of man?" The catechism answers that our ultimate purpose, the reason we were put on this earth, "is to glorify God, and to enjoy him forever." What an awesome responsibility and privilege! Glorifying God and enjoying him ought to pervade every part of our daily lives, including our professional and cultural activities. Like the stones ready to cry out in Luke 19:40, our engineering, science, mathematics, and technology are ready to cry out in worship of the Almighty. In this section, we will look at a few ways we can glorify God with technology and a few ways we might enjoy God with technology.

Technology can help us glorify God in both formal and informal worship settings. Attend a church service almost anywhere in North America or Europe and you'll spot technology all around you. Used appropriately, it can enhance worship. Some uses of technology are rather conspicuous: reading the Bible on a smartphone, singing a hymn with words projected on a screen, watching a live report from mission-aries in the field via a web conferencing tool, or connecting worshipers in two venues via video link. Many churches also use very noticeable technology for a church webpage or an email prayer chain.

Less obvious are the background technologies that churches use, which we hardly notice anymore but that contribute to authentic, sincere worship nonetheless. The heating and air conditioning of the building provide a comfortable temperature that allows us to focus. Electric lights allow us to see inside even on dark days. An elevator allows those with trouble walking to avoid the steps in a multilevel church building. A sound system lets us hear the preacher and the music in a large auditorium. Even the musical instruments played during worship are technological devices. Some congregations serve grape juice at Communion, making use of a technological invention by Thomas Bramwell Welch in 1869 that prevents the juice from spoiling or from fermenting into wine.

Technology can also detract from worship. As a tool, it should be as transparent as possible so that it allows us to focus on loving God, not the device. For example, most of us do not even notice the light fixtures in our worship spaces; they simply help us see. However, even lights can be used irresponsibly, such as spotlights that unduly focus our attention on a performer instead of worship, or dramatic laser light shows that become distracting entertainment. We regularly and intentionally use technology to amplify certain abilities, but we must be wary of what abilities it might narrow. We need to discern how best to deploy tech-nology and when to refrain from using it.

Our worship does not start and stop with the formal service in a church building. The church is the body of believers, and worship can and should be an ever-present mindset and continuous act. Technology can also aid our less formal worship. For example, despite the dangers of social media, it can be used to encourage fellow believers and connect with seekers. Facebook posts are a form of worship if they share Scripture, encourage fellow believers, provide a snippet of personal testimony, or wonder at the beauty of creation through some stunning pictures of nature. Connecting with family in a distant location through texts or live video chats can be a form of worship when we pray for one another and provide spiritual encouragement. One can even worship by browsing Wikipedia and praising God for the wonders of the natural world and the diversity of our culture.

Technology can also help us enjoy God. For example, engineering work depends on the laws of physics that do not change arbitrarily. We can count on those natural laws because the Holy Spirit providentially upholds them. We can appreciate God the Father as Creator when we exercise our creativity by designing and inventing. We are made in the image of God, and our creativity is part of that reflection. Telescopes let us see the wonder of the deep heavens. Satellites let us see the precious orb we live on. X-ray and MRI machines let us see the secret beauty and incredible complexity inside the human body. Through all these sights we can appreciate and enjoy God.

THE CULTURAL MANDATE

The commands to love God and neighbor are central to Scripture, but there is another command very early in the Bible that we should also consider in relation to technology: the command to create culture.

Engineers love to create. We have a creative muse that rarely stops singing, spinning new designs in our heads, new solutions for problems we encounter, new ideas for situations we notice. Software engineers love to write a new program, debug it, and then watch it work. They love tinkering with software to add a new feature or simply refactor the existing code into something more elegant. Once we crank out a design, we look for something novel and new. We regularly spot inefficiency or opportunity for enhancement and pounce on it.

Engineers share this creative gift with artists, musicians, writers, comedians, and many other professions. The medium and content of our creativity varies greatly, but they are all rooted in a common human characteristic inherited as part of the *imago Dei*. Creativity is fundamental to what it means to be human. It is also an essential gift that allows us to carry out our responsibilities as God's appointed stewards over his creation. The very first assignment—the original commandment—that God gives to humans is found in Genesis 1:28, "God blessed them and said to them, 'Be fruitful and increase in number; fill the earth and subdue it. Rule over the fish in the sea and the birds in the sky and over every living creature that moves on the ground.'" The command to be fruitful is not only about increasing in number by getting married and having children. It is also figurative. As defined in chapter two, the cultural mandate is God's first commission to humankind, a directive for us to be fruitful by cultivating and caring for the earth in new ways, including creating technology. We honor God the Creator by taking the materials that he has entrusted to us in the creation and multiplying them into greater diversity and beauty.

Creating culture is not a frill, a luxury, or a form of recreation. It is a matter of obedience. "If Christ is the creator of everything, then we must realize that his lordship is as wide as creation. Nothing in this universe escapes his lordship. And if his lordship is as wide as creation, then our obedience to his lordship must be as wide as culture."[6] Creating culture in diverse forms honors God. Composing music honors God; writing poetry honors God; designing technological devices honors God. Sin may distort these cultural artifacts, staining that honor, but we should not conclude that we must then abstain from all activity because it cannot be perfected. Rather, we seek God's grace as we humbly pursue the work of cultivating his creation.

Earlier in this chapter we pointed to technology as a means to an end: an aid to love God, to love our neighbor, and to disseminate the good news. In these situations, technology is a tool to help us reach a goal—it is not an end in itself. However, there is a biblical way that technology design can become an end. Exercising the cultural mandate is in itself a way to honor and glorify God. Technology is a cultural artifact; invention and engineering are cultural activities. We find joy and delight in the creation of the tool—even if, and perhaps especially when, we design it for someone else. It is a gift to find a particularly elegant solution while developing software, designing a bridge, inventing a new thermal deposition process, or doing a myriad of diverse engineering activities. The eureka moment that comes as we recognize a particular beauty within a design is part of our fulfillment of the creation mandate, such as the beauty seen from the roadside in the Netherlands where the old technology of the windmill and the new technology of the wind turbine stand side by side (fig. 9.1). God recognizes its beauty too! When we design technology to honor God, we should give thanks to him for the talents he gave us and the resources he provided.

Figure 9.1. Windmill and wind turbine

CONCLUSION

Thanks be to God that we do not need to leave technical professions nor abstain from technology to follow Christ. The truth is just the opposite. Following the biblical commands highlighted in this chapter, we've seen that technology can help us obey the commandments to love God, to love our neighbor, to bring the gospel to the ends of the earth, and to create culture. We are, in fact, called to pursue technological professions, designing and using technology in ways that are faithful.

The examples in this chapter for how to do this are not meant to be limiting but to suggest possibilities. New technologies will bring new opportunities to pursue faithfulness. They will also present potential hurdles about which we must be discerning and prophetic. God's will for us revealed in Scripture is multilayered and challenging. What does God ask you to do with technology in relation to principles like justice, mercy, and grace? What joy does God call you to in designing new technologies for the sheer delight of creative production of new cultural tools?

This chapter provided a more personal examination of the connections between faith and careers in technology. The next chapter is personal and practical as well, looking at some of the real questions that Christian engineers have faced in their work.

CHAPTER TEN

LETTERS TO A YOUNG ENGINEER

*If we would have our creations be true, beautiful,
and good, we have to attend to our hearts.*[1]

FREDERICK P. BROOKS

We recognize that a field guide can be useful for sketching the general landscape, but one will always encounter novel situations in the real world. Derek Schuurman explores a few practical issues in the form of letters exchanged between a young engineer and his mentor (a former professor). These letters are examples of wrestling with and discerning how one practically lives out one's faith in the context of the "real world"—where the rubber meets the road.[2]

Not everyone who works as an engineer continues to be inspired by stirring dreams as described in our first chapter. The beige cubicle farm found in large corporate settings can sap creativity and reduce dreams to drudgery. The *Dilbert* comic strip paints a picture of a hapless and unhappy engineer who is compelled by a "pointy-haired boss" to develop meaningless and ill-fated technical projects.[3] Undoubtedly the *Dilbert* comic, along with TV shows like *The Office*, resonate because many engineers (and other office workers) find themselves working on projects for which they lack passion or which are poorly managed. This can lead to stress and even despair. Work can cease to feel like a calling, becoming a means to other ends like early retirement, donating to Christian causes, or simply paying the bills. A Christian worldview must be more than mere platitudes; it needs to inform the rough and tumble of everyday life, lived out in the real world where the "rubber meets the road."

In this chapter, I attempt to acknowledge the complex and nuanced "real world" challenges that will inevitably arise through a series of imaginary letters between a young engineer (Dan) and his former engineering professor at a Christian university (Professor van Wijs). The inspiration for this chapter arose from my own experiences and conversations with real Christian engineers and Christian college graduates who are grappling with what it means to be faithful in the real world. These letters do not resort to platitudes, nor are they overly prescriptive, but rather they attempt to illustrate how a wise mentor can provide helpful encouragement in the life of a young Christian engineer.

From: Daniel Prentice
To: Prof. van Wijs
Subject: Update from a former student

Dear Professor van Wijs,

How are things going? It's hard to believe it has been four years since graduation! As I shared in a previous letter, I moved to a town called Springfield on the west coast where I started my first job as a junior engineer at the ACME Corporation.

I really enjoyed my time studying engineering at a Christian college, but I am finding that work in the "real world" can be challenging. I sit in a large, beige, cubicle farm as part of a team that designs a part for a small sub-assembly that is used in larger products. I report to our team lead, who reports to our manager, who in turn reports to a senior manager, and so on. My team is responsible for a logic circuit that is used in a larger module implemented in a programmable logic chip, one of dozens on a circuit board, which is one of seven in a chassis, which is one of several used in aircraft avionics. Like the parts I help design, I feel like a small cog in a giant machine.

Frankly, there are some days at work when I see my life reflected in the comic character *Dilbert*. My supervisor is not "pointy haired" like the boss portrayed in the *Dilbert* comic, but he does share other traits. For instance, he does not seem to understand or appreciate the technical challenges our department faces. He often raises his voice in meetings, hovers critically over employees seated at their desks, imposes unreasonable project deadlines, and takes credit for the ideas of others.

My coworkers come from a wide variety of different backgrounds. Some are easy to work with and others are somewhat eccentric. My cubicle-mate, Larry, seems to be unaware of basic social norms. He periodically clips his toenails at his desk and frequently tells crude jokes. Things often get tense as a deadline looms, and people start deflecting blame to others, leading to conflict. The tight deadlines are coupled with expectations to put in long hours, and I often come home late, making it difficult to get involved in activities outside of work.

In college I was inspired by an exciting, comprehensive kingdom vision but, now that I am in the "real world," work often seems tedious, stressful, and inconsequential. Truth be told, I often wonder exactly how my faith really matters in a large corporate setting with little opportunity to make a real difference.

Anyway, I wanted to send you my new email address and take a moment to reconnect.

Best regards from your former student,

Dan

From: Prof. van Wijs
To: Daniel Prentice
Subject: Re: Update from a former student

Dear Dan,

What a delight to hear from you! As it turns out, I was hoping to get your new email address and reach out to ask how you are settling into a new home and job.

Some of the feelings you describe are not uncommon when making the transition from college to work. As you know, I worked as an engineer for several years before I began teaching, so some of your experiences sound familiar. I know firsthand that it's easy to become discouraged at times, especially when work feels inconsequential.

I recall when I began my first engineering job in a new city far from home. At that time a wise mentor pointed me to the words of Jeremiah to the Jewish exiles in Babylon who were longing for home. God told them to "build houses and settle down; plant gardens and eat what they produce. . . . Seek the peace and prosperity of the city" (Jeremiah 29:5-7). The call to build, plant, and eat, even in exile, are echoes of the original cultural mandate. Babylon was the high-tech center of their day, and like those exiles, we may find ourselves feeling like strangers in a strange land. And yet, just as they were called to seek the peace and prosperity of the city, we are called to seek the peace and prosperity of our communities and workplaces. Your work at ACME designing useful tools is a legitimate contribution that serves both God and neighbor.

But, as someone who spent years working in an engineering setting like yours, I do not want to dismiss your struggles with mere clichés and platitudes. I suspect that as Christians we can idealize the notion of vocation by emphasizing the meaning and impact of our work. However, the sense of being a "cog in a wheel" in a machine beyond your control is not an uncommon experience in large companies.

A cosmic kingdom vision is inspiring, but individually we are called to be salt and light in the places where we find ourselves. I admit that triumphant narratives are sometimes propagated through the stories in our alumni magazine, profiling people who built influential companies or forged high-impact nonprofit organizations. Our individual job is not to single-handedly change the world but rather to be a faithful part of a much larger "cloud of witnesses" spread throughout space and time.

Don't underestimate the importance of regularly praying for your workplace and for your coworkers (even your boss!) and asking that God will use you. Do your work diligently, ask good questions, listen closely, show empathy to coworkers (including Larry!) as well as to customers. Focus on your circle of influence, on how you interact with others, and the design decisions for which you are responsible.

Remember that your engineering job is just one sphere in which you are called to be faithful. Our primary calling is to be a disciple of Jesus Christ, but we have multiple secondary callings—not only our job, but also in our local church, with your family, your friendships, your neighborhood, and through volunteer opportunities. Your life and witness do not just consist of your paid work.

It is wonderful to hear from you! I will pray for you as you settle into your new life and work.

Stay in touch!

Sincerely,

Prof. van Wijs

From: Daniel Prentice
To: Prof. van Wijs
Subject: Re: Update from a former student

Dear Professor van Wijs,

Thanks for taking the time to reply, and for your prayers.

Your letter was a good reminder to keep things in perspective (many of your comments reminded me of the concepts covered in one of the books we read in our capstone course, *A Christian Field Guide to Technology for Engineers and Designers*). To be clear, I am extremely grateful for the expansive Christian worldview presented in my college education. However (and please don't get me wrong), sometimes these ideas feel like platitudes without much application to the rough and tumble of the corporate engineering world—especially when one is working at the bottom of the ladder.

What were your experiences like working in industry before you went into teaching? How did you find the challenges of working in the real world? (No offense, I suppose college is also the "real world," but it's a different world, if you know what I mean.)

Thanks and take care,

Dan

P. S. If you are ever in our area, please drop by for a visit, and we can grab a coffee. In my humble opinion—as someone who grew up in the Midwest—they really know how to make coffee out here on the West Coast!

From: Prof. van Wijs
To: Daniel Prentice
Subject: Re: Update from a former student

Dear Dan,

Thanks for your thoughtful note. I am also glad to hear that
you are enjoying the local coffee (although I think the "Providential Bean" coffee shop here on campus has the best dark
roast coffee I've ever tasted).

Your note about moving beyond platitudes is a point well
taken. The issues engineers face in the "real world" are
seldom trivial (and you are correct that industry is a much
different context than a Christian college). However, the fact
that you are thinking about the practical working out of your
faith as an engineer is an essential first step.

To be honest, I did not go to a Christian college, but rather I
attended a large, secular engineering school. To be sure, I was
equipped with a fine technical toolbox, but there was something missing in my education. As a young graduate, I was not
very familiar with a "comprehensive Christian worldview."
Frankly, at that stage in my life, I was quite enamored with
strictly technical pursuits and did not think too deeply about my
work. It was not until years later that I began to sense a disso-
nance in my life, and I began to seek connections between my
faith and my profession. At one point, I entertained becoming a
pastor or missionary since that felt like an easier way to
connect my faith with my vocation. But God had given me
technical talents, and I did not want to bury them.

Inevitably, I did encounter some thorny dilemmas in my
work, and I confess I often responded in a self-righteous and
moralistic way. Unfortunately, this posture created barriers to
helpful and meaningful interactions with my colleagues. It was

not until later that I met a wise Christian mentor who helped me to develop a more winsome approach to influencing my colleagues, not only for addressing ethical quandaries that arose but also as a witness to the gospel.

Although I had an impact on the workplace, it also had an influence on me. In hindsight, there were several patterns that I observed in the engineering workplace that are common and can misshape you. For example, many engineering cultures demand long hours, placing work ahead of friends and family and your own physical and spiritual well-being. Engineers are often motivated to work long hours to gain the approval of coworkers or as a means of affirming self-worth. Living without proper regard to rest and sabbath has consequences. Moreover, engineers are often paid well for their skills, and so you will need to be on guard against greed and a sense of self-reliance. Finally, while working on successful, high-profile projects is satisfying, it can also lead to pride. Don't let these patterns shape you like the seed that fell among the thorns in the parable of the sower—seed that is "choked by life's worries, riches and pleasures, and they do not mature" (Luke 8:14).

At the outset of our careers, we may not be inclined to any of these distorting patterns. However, the workplace can influence us insidiously, shaping our hearts to rival visions of the good life, often without even realizing it. The words of the philosopher Søren Kierkegaard come to mind: "The greatest hazard of all, losing one's self, can occur very quietly in the world, as if it were nothing at all."[4]

The antidote to "losing one's self" to rival visions of the good life is to cultivate regular spiritual counterpractices aimed at keeping our hearts set on Christ. Don't forget the practice of regular Scripture reading, and pray that God will equip you through his Holy Spirit. We cannot expect to be effective in the kingdom if we don't maintain a relationship with the King!

Try not to limit your reading to technical manuals and engineering magazines, but seek thoughtful books by wise Christians, both contemporary authors as well as authors from the Christian past. As you gain more experience, be attentive to God's leading as new opportunities for work and service may present themselves. Seek fellowship and encouragement by connecting with other Christian engineers and wise mentors (I can attest to the importance this has been in my own life).

After nearly a decade in industry and after much prayer, I felt a call to leave industry and enter teaching. Please don't interpret this as advocating that one needs to *leave* engineering in order to serve God. This was the path I felt led to follow, and your path will likely take you to different places. Either way, God's word provides a lamp for our feet (Psalm 119:105) and he will help you discern your path as you seek to use your gifts and talents in his service.

Grace and peace to you,

Prof. van Wijs

From: Daniel Prentice
To: Prof. van Wijs
Subject: Re: Update from a former student

Dear Professor van Wijs,

Thanks for sharing some of your experiences. To be honest, my questions are not just academic, they were prompted by some specific situations that have recently come up at my work.

We are nearing a delivery deadline on a large project, and it's clear that our product is missing features and needs more testing. In fact, I sensed that the original deadline was far too optimistic right from the beginning. I can't tell you much about the product since I signed a nondisclosure agreement (NDA) that prohibits me from talking about the details.

Although the deadline remains a few weeks away, our sales team has promised the customer a firm shipping date. I am now hearing rumors that the sales team is pushing to meet that deadline regardless of whether the product is ready. I am just a junior engineer, so I have little say about the matter, but I still feel responsible for the design.

Another situation has also come up. Because of the tight schedule, we are now being pressured to work on Sundays as the project deadline approaches. I am just getting settled into a local church where I have joined a small group and have started to get more involved. I am playing the bass guitar on Sunday mornings and helping out with the youth group. I am uncomfortable with the prospect of working Sundays, but I don't want to disappoint my boss and coworkers.

Finally, I am working on a smaller project on the side, and one of the project managers has discreetly asked me to adjust my project time sheet by entering more hours than I actually worked on it. The reason he gave me is that this project requires additional equipment costs that need to be covered. I am not comfortable doing this even though the project costs have actually increased. It seems dishonest to inflate the hours I report, but it's what I have been asked to do.

Sorry for bothering you with all these issues. I recall we discussed ethics in one of our classes and read various case studies, but somehow those case studies seemed so remote at the time. Would you have any practical advice for how to navigate these situations?

Thanks,

Dan

From: Prof. van Wijs
To: Daniel Prentice
Subject: Re: Update from a former student

Hi Dan,

To be honest, your previous email made me wonder if there wasn't something more behind your questions, but I was reluctant to pry into your situation. Thanks for the privilege of sharing these challenges with me. I am also glad to hear you have found a home church and have started to get involved! Here are some of my thoughts regarding the situations you have shared.

The issue surrounding tight deadlines is not uncommon but hard to judge without knowing more about the context of your situation. I am glad you are respecting the nondisclosure agreement you signed (and I trust that the company is using an agreement that is both fair and reasonable). You will need to discern the situation in order to determine what the appropriate action is in this case. What I can offer are some general thoughts for you to consider.

In general, schedules are an area where you will need to cultivate trust by being honest with managers and customers. You should let people know about potential schedule slippages with your project as they become apparent, rather than waiting until the deadline is imminent. That being said, estimating project timelines is something that takes experience. By the way, I would recommend reading more about project management, even though you are not a project manager. Take time to observe how projects unfold and how to manage risk. Cultivate the skill of accurately estimating effort and time, something that will help you to make realistic time commitments and develop your reputation for being honest and trustworthy.

One important question is what you mean when you say the product is not ready for shipping. Is the product missing some nonessential features, or does it lack testing, or are there bugs and faults in the system? If the product is working but not feature complete, the salespeople could negotiate in good faith with the customer who may be willing to accept the product with missing features to receive it on time.

However, if the product is faulty or not properly tested, that is another matter. The stakes are even higher if a product design is one on which human lives and well-being depend (and since you work in avionics, I can only speculate that this might be the case). Never forget that as an engineer you have a *fiduciary responsibility*, which is a legal requirement to "hold paramount the safety, health, and welfare of the public." Many industries have applicable standards for testing which are established by regulating bodies. If the product does not comply with the required standards, or if it is putting public safety at risk, you are obliged to say something.

If you find yourself in a situation where prudence and justice require you to speak out, you will need to exercise the virtue of courage to do so. In rare cases, you may need to muster the courage to do what needs to be done even though it may cost you your job. However, there are other ways to proceed first. There is wisdom in following a pattern like the approach outlined in Matthew 18 for settling disputes. First talk with your coworkers about the situation and together bring your concerns to your team leader. If potential safety issues are not addressed, then go to your manager. If a product poses significant danger, it may warrant going further up the chain if the manager does not listen. In the avionics industry, there will be best practices, standards, and regula-tions to which you can appeal that govern the process by which products are designed and brought to market. Do not shy away from asking questions about safety, but also be prepared to learn from more experienced engineers about best practices and what safety margins are appropriate in your field.

Moreover, as Christians, we are called to love our neighbor (which includes your customer). This is where designs can be influenced by consideration of the various design norms. Do you recall the design norms we covered in your classes? These include things like cultural appropriateness, clarity (transparency), stewardship, harmony (aesthetics), justice, caring, and faithfulness. One way to show love and care for your neighbor is to consider and advocate for these design considerations as your influence allows. As you garner the respect and trust of your colleagues, your voice will gradually carry more influence during design reviews.

The issue of working on Sundays is becoming a more common situation for Christians who navigate our 24-7 working world. Since regular spiritual practices (such as Sunday worship) are crucial to equip you in your life and work, you may want to push back against demands that undermine your ability to practice these. I am reminded of Daniel in Babylon who continued to observe Israelite dietary laws and prayed to God three times a day, despite the pressure to do otherwise. I think we should normally strive to worship with God's people whenever we can (Hebrews 10:25). Granted, there are some jobs that are considered "essential," and require people to legitimately work on Sundays. I don't know all the details of your situation, but it sounds like this might be a case where working on Sundays arises from poor project management rather than providing an essential service. When I worked in industry, I occasionally worked overtime on evenings and Saturdays during peak times, but with the understanding that

I was not available Sundays. My boss was usually understanding, but your situation may differ. Avoid being adversarial; are there opportunities to creatively negotiate alternatives? Are there ways you can judiciously work extra hours at other times to contribute to the project while still keeping Sundays set aside for worship? These discussions need to be friendly and diplomatic. Even so, if the culture at your company is one that pushes excessive overtime, you will need to discern how to maintain a healthy work-life balance. Approaching your manager on this topic will require that you exercise the virtue of courage, but having a track record as a faithful worker and diligent team player will make these conversations easier. At the end of the day, if a company habitually requires hours for nonessential work which prevents regular worship and family time and they are inflexible about it, you may be justified in looking elsewhere for employment.

The last situation you mention regarding your timesheets seems more straightforward. The Bible speaks plainly against false weights and measures (Proverbs 20:23), and reporting false times is a type of false measure. In addition, one of the fundamental rules of the NSPE (National Society of Professional Engineers) Code of Ethics is to avoid deceptive acts. Since your timesheet is used to bill the customer, you should not report your hours deceptively. Justice demands that your customer be billed for the proper number of hours. However, if there were additional costs in another area, these should be communicated and negotiated in good faith with the customer. I would speak gently and respectfully to the project manager to tell him or her that you are not willing to enter false information on time sheets and why. This will also require courage.

Discerning how to respond in various situations ultimately requires wisdom, and so it is important to pray for discernment and for the Holy Spirit to equip you.

Sorry for the long-winded reply and for not providing more straightforward answers. What I will do is pray that God will equip you to become the kind of person who can wisely discern how to faithfully respond to these situations and to others that will undoubtedly arise. In the meantime, do what you can to become a respected voice and cultivate good relationships in the workplace—it will help when challenges inevitably arise. Remember the words of Romans 12:2: "Do not conform to the pattern of this world, but be transformed by the renewing of your mind. Then you will be able to test and approve what God's will is—his good, pleasing and perfect will."

Sincerely,

Prof. van Wijs

P. S. In case you hadn't heard, there has been a lot of excitement on campus—our college basketball team just made it to the finals last night!

From: Daniel Prentice
To: Prof. van Wijs
Subject: Re: Update from a former student

Dear Professor van Wijs,

Thanks for your letter and your advice. Glad to hear the basketball team made the finals! I played basketball for one season at the college but spent most of my time warming the bench. I am hoping that once things settle down at work I can find others to play some pick-up basketball on the weekends.

By the way, I have some updates on my work situation: we have solved several of the bugs that plagued our project, but more work needs to be done. I have offered to work overtime as the deadline approaches, but on Saturdays, not Sundays. For now, the project leader is fine with this arrangement, but we will see what happens as the deadline gets closer.

Thankfully, the timesheet issue is no longer relevant—the side project was shelved to help us to focus on delivering the bigger project. However, if the issue comes up again, I have resolved to gently and respectfully push back against any pressure to tinker with time sheets.

Thanks again for your advice. I wish I could simply open my Bible for the answers to navigate situations in my life, but it is not always that simple. Incidentally, your reference to the story of Daniel caught my eye, not only because I share the same name, but the book of Daniel has come up in recent sermons at church. I am encouraged by the story of Daniel and have started reading through the book of Daniel for my own devotions.

Thanks again for your advice—

Dan

From: Prof. van Wijs
To: Daniel Prentice
Subject: Re: Update from a former student

Dear Dan,

Thanks for the note—I'm glad your project is progressing and you have begun to discern your way forward!

I think you will find the book of Daniel helpful. Daniel is a good example of someone who strove to live faithfully in Babylon, the "high-tech" center of his day. While he adapted to his situation, he also remained faithful to God's law. Following Israelite dietary restrictions is an interesting example. When the king's food and wine were presented, he did not protest or start a hunger strike, but rather gently asked the guard to allow them to eat vegetables and water. The guard was fearful that he would not look as healthy as the other young people their age and that he would then be held responsible. In reply to this Daniel asks that he wait ten days and then compare his appearance to others eating the king's food and wine. Daniel's approach illustrates both obedience and trust. We read that the Lord "caused the official to show favor and compassion to Daniel" and that after the ten-day test "they looked healthier and better nourished than any of the young men who ate the royal food" (Daniel 1:9, 15). Later, we read about how Daniel resolved to remain faithful in praying three times a day (a practice that lands him in the lion's den). Likewise, Daniel's colleagues Shadrach, Meshach, and Abednego faithfully refuse to kneel to King Nebuchadnez-zar's image of gold, even with the threat of being thrown into a fiery furnace. In their reply to Nebuchadnezzar, they declare that while God is able to deliver them from the blazing furnace, *even if he does not*, they will not bow down to the statue (Daniel 3:16-18). In all these events we see that Daniel and his friends in Babylon required courage to remain faithful, but that the hand of the Lord was active behind the scenes.

The story of Daniel provides a helpful contrast to the *Dilbert* comic strips you mentioned earlier. The world of Dilbert is one of despair and absurdity but, if we seek to honor God in the places he has placed us, we can strive to be faithful while leaving the outcome to him. This is true if we are promoted to high positions like Daniel, but it is also true for those who remain faithful engineers working in relative obscurity in the context of a large corporation. If our work is done for God then "your labor in the Lord is not in vain" (1 Corinthians 15:58). Either way, strive to be a Daniel rather than a Dilbert![5]

The story of Daniel is helpful, but all of Scripture is "useful for teaching, rebuking, correcting and training in righteousness, so that the servant of God may be thoroughly equipped for every good work" (2 Timothy 3:16-17). Following our own hearts we will tend to stray from the right path, but God's word is a "lamp for my feet" that can keep us on track. This is why the practice of regular Bible reading and prayer is essential. I know firsthand that time pressures in industry (and even in teaching) can often squeeze out regular devotions, so it's encouraging to know you are making this a part of your daily life. Keep it up!

Don't forget that we live for an audience of one; we live *coram Deo*, before the face of God. Our entire life is a response to God, and our response is either one of obedience or disobedience. If Christ's lordship extends over all of life, then his lordship must also extend to engineering and technology. In the words of the late professor Lewis Smedes, we are called to "go into the world and create some imperfect models of the good world to come."[6]

Needless to say, there is much more that could be said about engineering and the biblical story (and I am glad to teach at an institution where I can freely do that!). And remember that faithful engineering is not only a matter of the mind and intellect, but also of the heart. "Above all else, guard your heart, for everything you do flows from it" (Proverbs 4:23).

While spiritual formation and studying the Scriptures were part of your Christian university education, the truth is that you never graduate from it—it's a lifelong project (also for your professors)!

Take care,

Prof. van Wijs

From: Daniel Prentice
To: Prof. van Wijs
Subject: Re: Update from a former student

Dear Professor van Wijs,

Thanks for the reply. I have one more matter that has been on my mind.

Our pastor has recently challenged us to share our faith more openly—also in the workplace. I have attempted to share my faith in the lunchroom with my coworkers on multiple occasions, but unfortunately, my feeble attempts at presenting the gospel have been stilted and awkward. Do you have any experiences sharing your faith in the workplace or any practical advice? After all, being a Christian engineer ought to be evident in more than just how we design.

I am planning to be back on campus sometime soon, so perhaps we can meet over a coffee at the Providential Bean coffee shop on campus (although I still maintain that the coffee here on the West Coast is much better than anything in the Midwest).

Take care,

Dan

From: Prof. van Wijs
To: Daniel Prentice
Subject: Re: Update from a former student

Dear Dan,

Thanks for your thoughtful question. You are absolutely correct—living the gospel must be evident in more than just how we design.

I confess that as a young engineer I was quite shy about my faith. I suppose I wanted to gain the respect of my coworkers and was reluctant to "push" my faith on others. Later, I initiated some awkward attempts to share my faith that fell flat. Remember that wherever you work, God is already there.

Gradually, over time, I discovered that as I got to know my coworkers the issues of life would inevitably come up in conversation, whether it was the lunchroom, over a beer after work, or while driving to a customer site. Even an innocuous social event, like going out to watch a newly released science fiction film, would lead to questions like, Will computers ever become conscious? What does it mean to be human? Where is technology headed? I found that developing a relationship with my coworkers and listening to them gave me the opportunity to respectfully share my own story. In winsomely sharing my own hopes and fears my faith would naturally come up as a part of that story.

How do your coworkers see you? Do your coworkers see you as a joyful, trustworthy person of integrity? Don't underestimate how your life and actions can also be a witness. Interestingly, my practice of not working on Sundays became a springboard for also talking about my faith. At one point I invited a coworker (who has since become a close friend) to accompany me to an Alpha program at my church. Don't forget the words of 1 Peter 3:15: "Always be prepared to give an answer to everyone who asks you to give the reason for the hope that you have. But do this with gentleness and respect."

I pray that God will bless you in your work and witness for him. Thankfully, even when we stumble or when we feel our work seems inconsequential, we can be assured that our world belongs to God and that our salvation rests on Christ alone.

May God continue to bless you and keep you in your work and may he "equip you with everything good for doing his will" (Hebrews 13:21).

Sincerely,

Prof. van Wijs

From: Daniel Prentice
To: Prof. van Wijs
Subject: Re: Update from a former student

Hi Dr. van Wijs,

I just booked some tickets and will be travelling back to campus for the homecoming weekend next week! I hope to catch a basketball game and visit with some of my college friends. If time permits, let's meet Friday afternoon at the Providential Bean over a coffee.

Thanks again for your encouragement. I am looking forward to catching up!

Take care,

Dan

P. S. I'll be sure to bring a bag of my favorite West Coast coffee beans with me as empirical evidence of how good the coffee is here!

QUESTIONS FOR REFLECTION OR DISCUSSION

The following are some reflection questions for discussion based on each of the chapters in this book. To continue the conversation with related links, articles, and resources, visit the website associated with this book, www.ivpress.com/fieldguide.

1 DREAMS TAKE FLIGHT

1. Have you ever built something that started from a yearning or dream of a better way? In what ways did the dream shape the final result?

2. Consider a technology that you use regularly. What dreams may have inspired it?

3. Name some examples where technology dreams turned into nightmares. Where did they go wrong?

4. What are three new technologies that have changed your life the most? Did they change your life for better or worse? Why?

5. Do you have a yearning for a technological device to solve a particular problem or enable a certain possibility? Explain your dream and why it is significant to you.

6. What dreams do you have for your work in technology? If you are currently studying in a technology-related discipline, what are your dreams for your future work? If you are currently working in a technical area, how have your dreams shaped your path?

7. How might our faith and the biblical story shape our dreams, yearning, and imagination?

2 A SURVEY OF TECHNOLOGY
AND THE BIBLICAL STORY

1. When Jesus summarizes his entire mission in Luke 4:18-19, he does so by referring to the seven-day rhythm of creation, that is, the culmination of the sabbath cycle in the Year of Jubilee. When Jesus says he is proclaiming the Year of Jubilee, what implications does this have for us who seek to be imitators of Christ in our creating and shaping of technology?

2. How does Colossians 1:17 reflect the most succinct and yet comprehensive creation account? What implications does this passage have for technology?

3. How does our dependence on God impact how we work with technology?

4. In Genesis 2:15, the phrase most often translated as "put him in the garden" can also be literally translated "rested him in the garden." How do the important biblical words *rest*, *work*, and *care* (in Genesis 2:15) come together to guide our understanding of biblical stewardship as it pertains to technology?

5. Explain how idolatry and our fall into sin go hand in hand. Provide a contemporary technological example of idol making and idol worship.

6. How is the redemptive work of Jesus Christ told from Genesis to Revelation? What implications does this have for the nonhuman creation? For our technology?

7. Seeing the work of the Holy Spirit in and through creation from Genesis to Revelation is essential for followers of the risen Lord. How do we listen, wait, and cultivate space in our technological world for the work of the Holy Spirit?

3 FIELD RESPONSIBILITY

1. Is there any technology (device or product) in your home that you have used in ways the inventor probably did not intend? Was it a successful and good use?

2. To what extent should we hold the designer of a technology responsible for injury caused by unanticipated use of the product? How much should we expect an engineer to anticipate?

3. Tradeoffs are an inherent part of designing technology. When might the benefits outweigh any detriments when making a tradeoff between different design goals? That is, what ends might justify the means?

4. Why is it so common after an accident for those involved to claim, "I was simply doing my job," as an excuse that they hope absolves them of responsibility?

5. Think of a technology used at your church, perhaps during the worship service. Does the technology aid and support worship in spirit and truth? Does it enhance the worshiper's dialogue with God? In what ways does it detract from worship? How might we avoid such distractions?

6. Choose a common household technology and brainstorm alternative uses for it—both serious and humorous. What biases do you discern as you reverse engineer its purpose? How do different models of appliances display different biases?

7. Typical decision criteria that we use to evaluate design alternatives are often easily quantified, such as cost, weight, and reliability. Broader criteria such as safety, environmental impact, or justice are more difficult to quantify, but are important considerations. How would you put a number on these in ways that would best lead to a wise design decision?

4 FAITHFUL DESIGN GUIDES

1. Consider how each of the norms could be applied in the design of a new office building as part of a revitalization project in an older downtown area. Could any tensions arise when striving to honor all the norms?

2. Throughout Scripture, God makes it clear that he has a special concern for the poor, the widow, and the stranger—those that are less well off. How can technology be directed toward justice, caring, and love of the disadvantaged when the potential for financial return so often drives technological decisions?

3. You may have heard that sometimes well-intentioned short-term mission trips end up doing more harm than good for the communities they visit. Why do you think that might be? How might attention to design norms help prevent such an unintended result?

4. Engineers, especially early in their careers, may have very little influence over the choices and directions for the technology projects to which they are assigned. What can they do in such situations to honor design norms?

5. "The heavens declare the glory of God" (Psalm 19:1). Can technology also declare God's glory? Which norms suggest this? How might we encourage it?

6. This chapter suggested that one way to incorporate design norms into a design process would be to include them in a decision matrix. However, equating norms to technical design criteria might minimize the norms, making it easier to rationalize giving them less priority. What are some other ways that the norms could be systematically considered in the technology design process?

5 BEYOND ENGINEERING ETHICS

1. Have you ever encountered any ethical challenges in your life or work? How did you deal with them? How did your faith inform your actions?

2. Review the NSPE code of ethics. What ethical frameworks inform each section of the code of ethics: deontological ethics, consequentialism, or virtue ethics?

3. What norms do you see represented in the NSPE code of ethics? What norms are missing?

4. Can Christians embrace the NSPE code of ethics? What might Christians add to the code?

5. "Sound ethics and good business go together, for the most part and in the long run."[1] Do you agree with this statement? Can you cite personal examples or counterexamples? Have you ever experienced any tension between profit and ethics, and if so, how did you deal with it?

6. What ethical responsibilities do users of technology have? How do these compare and contrast to the responsibilities of those who design technology?

6 MODERN TOWERS OF BABEL

1. What do you think about the suggestion that technology is a result of the fall?

2. Have you detected examples of *technicism* in books, magazines, or TV shows?

3. In what ways can technology become an idol? Can you think of any modern-day "towers of Babel"?

4. How might a Christian engineer advocate for various design norms in a secular setting? How can a Christian be a winsome witness in these situations?

5. Have you been involved in any failed engineering or technology-related projects? What was the cause? Did sin, human finiteness, or both play a part?

6. Perform a "liturgical audit" of some devices in your life. How do they shape your daily habits and rituals? How might these habits and rituals shape your heart over time?

7 MOVING FORWARD BY LOOKING BACK

1. Henry Ford was once quoted (and requoted in dystopian novels such as *Brave New World* by Aldous Huxley) as saying, "History is bunk." Briefly explore the story of Henry Ford and Fordism. How is Ford's obsession with what he deemed the future and his disregard for the past reflected in how he made and developed the automobile? If Henry Ford had allowed Scripture and history to illuminate his technological vision, how might Fordist systems (i.e., Taylorism) and Fordist technology have looked different?

2. What criteria do we use to define progress today? How does this coincide with or diverge from biblical metrics we might use that would frame progress? Illustrate using a contemporary innovation.

3. For further research and exploration, consider the decline of railroads and the rise of the interstate highway system. What were the social, cultural, economic, and political forces that influenced this change?

4. Can you think of other technologies that were abandoned years ago but that have experienced a resurgence? What factors played a role in this decline, abandonment, or readoption?

5. In almost every generation, competing technologies vie for the same audience or claim to solve the same problem. What are the competing technologies of your generation? Step back a bit and identify the cultural forces that are driving the competition. What human wants and desires are at play?

6. Name one or two technologies introduced in your lifetime that have had the most significant impact on your daily life. What do you think your parents would identify as significant in their lifetimes? Your grandparents? Why do you consider them to be significant? How does this reflect your presupposed metric of progress?

8 TECHNOLOGY AND THE FUTURE

1. Think of your favorite science fiction book or movie. What view of the future does it portray?

2. Can you name any science fiction movies that paint a utopian view of the future? Why do you suppose there are so few movies of this genre?

3. What do you struggle with more, the pitfalls of technological optimism or pessimism? Do you think that working with technology makes one more optimistic or pessimistic about technology?

4. Do you think that transhumanism and Christianity are compatible? Why?

5. Do you agree with the authors that there will be technology in the new heavens and earth? Why or why not?

9 MUST WE LEAVE OUR NEURAL NETS TO FOLLOW HIM?

1. Identify two or three subtle technologies that are used in your church that enhance worship but that most people hardly realize are there.

2. What are some characteristics of professionals in technical disciplines that might make witnessing easier or more difficult?

3. Consider how we might demonstrate holiness more clearly with the following specific technologies: automobile, smartphone, lawnmower.

4. How do we care for our neighbor through technology if our neighbor does not have access to the high-tech tools necessary to truly heal their illness?

5. Consider how a few of the design norms or design virtues from the previous chapters, might help us properly design technology that helps us love God.

10 LETTERS TO A YOUNG ENGINEER

1. Does our work always have to be meaningful? How would you encourage a Christian engineer who complains that their work makes them feel like a "Dilbert"?

2. Have you encountered any ethical challenges like the ones described by Dan? How did you deal with them?

3. Do you share Professor van Wijs's view about working on Sundays? Why or why not?

4. Have you ever shared your faith with a coworker or acquaintance? How did you approach this and how was it received?

5. What Christian practices do you cultivate in your life? How might our dreams (as discussed in chapter one) be shaped by our spiritual practices?

6. If you had a mentor like Professor van Wijs, what would you like to ask him or her?

NOTES

PREFACE

[1]Neil Postman, *Technopoly: The Surrender of Culture to Technology* (New York: Vintage, 1993), 5.

[2]Corina R. Kaul, Kimberly A. Hardin, and A. Alexander Beaujean, "Predicting Faculty Integration of Faith and Learning," *Christian Higher Education* 16, no. 3 (2017): 172-87.

1 DREAMS TAKE FLIGHT

[1]Samuel Florman, *The Existential Pleasures of Engineering* (New York: St. Martin's Griffin, 1996), 125-26.

[2]This recollection is told by Steve VanderLeest.

[3]Heather S. Deiss and Denise Miller, "Who Was Katherine Johnson?," NASA, February 4, 2020, https://www.nasa.gov/audience/forstudents/k-4/stories/nasa-knows/who-was -katherine-johnson-k4.

[4]Michael West, "AstroAlert: Katherine Johnson, NASA Mathematician Who Broke Barriers, Has Died," *Lowell Observatory News and Highlights* (blog), February 27, 2020, https://lowell .edu/astroalert-katherine-johnson-nasa-mathematician-who-broke-barriers-has-died/.

[5]Abraham Kuyper, *Near Unto God*, adapted by James C. Schaap (Grand Rapids, MI: CRC Publications, 1997), 26.

[6]Kuyper, *Near Unto God*, 27.

2 A SURVEY OF TECHNOLOGY AND THE BIBLICAL STORY

[1]Eugene Peterson, *Leap Over a Wall: Earthly Spirituality for Everyday Christians* (New York: HarperOne, 1997), 3.

[2]Nicholas Wolterstorff, *Until Justice and Peace Embrace* (Grand Rapids, MI: Eerdmans, 1983), 70.

[3]James Skillen, *God's Sabbath with Creation: Vocations Fulfilled, the Glory Unveiled* (Eugene, OR: Wipf & Stock, 2019), 14.

[4]For creation accounts throughout Scripture, see Genesis 1–2; Psalm 8; Psalm 119:89-96; Job 38–41; John 1:1-14; Colossians 1; Acts 17:16-31; Hebrews 1; and Revelation 4.

[5]Ruth Schwartz Cowan, *A Social History of American Technology*, 2nd ed. (New York: Oxford University Press, 2018), 108.

[6]Skillen, *God's Sabbath*, 43.

[7]John Dyer, *From the Garden to the City: The Redeeming and Corrupting Power of Technology* (Grand Rapids, MI: Kregel, 2011), 45.

[8]John Calvin, *Institutes of the Christian Religion*, ed. John T. McNeill, trans. Ford Lewis Battles, (Philadelphia: Westminster Press, 1960), 53.

[9]Nancy Pearcey and Charles Thaxton, *The Soul of Science: Christian Faith and Natural Philosophy* (Wheaton, IL: Crossway, 1994), 35.

[10]Saint Augustine, *City of God*, trans. William Babcock (New York: New City Press, 2012), 539.

[11]Andy Crouch, *Culture Making: Rediscovering Our Creative Calling* (Downers Grove, IL: InterVarsity Press, 2008), 17-36.

[12]James Skillen, *God's Sabbath*, 61.

[13]Tim Keller, *The Prodigal God* (New York: Dutton, 2008). Authors such as Keller have emphasized that this trilogy of parables (lost sheep, lost coin, and lost son) are about God. All parables highlight a God who is continually "extravagant" and "over the top" in his grace to the world in which he loves.

[14]Roy Clouser, *The Myth of Religious Neutrality: An Essay on the Role of Religious Belief in Theories*, rev. ed. (Notre Dame, IN: University of Notre Dame Press, 2005), 241.

[15]Stephen V. Monsma, ed., *Responsible Technology* (Grand Rapids, MI: Eerdmans, 1986), 58-76.

[16]For summary accounts of our fallenness and its effect on creation throughout Scripture, see Genesis 3–6; Psalm 115; Ecclesiastes 1–2; Isaiah 44; Romans 1, 8:18-25; and Revelation 17–18.

[17]Roy Clouser, *Myth of Religious Neutrality*, 180-83.

[18]This term describes our modern tendency to break everything down to a single essence or binary duality. See Albert Borgmann, *Power Failure: Christianity in the Culture of Technology* (Grand Rapids, MI: Brazos, 2003), for his discussion of our culture of transparency and control.

[19]Roy Clouser, *Myth of Religious Neutrality*, 22.

[20]A helpful framework for understanding created goodness alongside broken humanity is the recognition that all created things possess a good *structure* that humans can shape and form in an obedient or disobedient *direction*. For additional examples on how structure and direction can be discerned, see Albert Wolters, *Creation Regained*, (Grand Rapids, MI: Eerdmans, 1985), 72-95.

[21]Heidelberg Catechism, Q/A 1.

[22]For redemption and restoration narratives throughout Scripture, see Isaiah 55; 2 Corinthians 5; John 3; 2 Peter 3; Matthew 24 and 28.

[23]Skillen, *God's Sabbath*, 117.

[24]From the Latin hymn *Veni Creator Spiritus*, ninth century. English paraphrase by John Dryden, 1693. Appears in numerous hymnals such as *Lift Up Your Hearts: Psalms, Hymns, and Spiritual Songs* (Grand Rapids, MI: Faith Alive Resources, 2013), #523.

3 FIELD RESPONSIBILITY

[1]Richard R. Galladertz, *Transforming Our Days: Finding God Amid the Noise of Modern Life* (Liguori, MO: Liguori Publications, 2009), 18.

[2]Robert A. Caro, *The Power Broker: Robert Moses and the Fall of New York* (New York: Knopf, 1974), 318-19, 335, 546.

[3]Neil Postman, *Technopoly: The Surrender of Culture to Technology* (New York: Knopf, 1992), 13.

[4]Nicholas Wolterstorff, *Until Justice and Peace Embrace* (Grand Rapids, MI: Eerdmans, 1983), 128.

[5]Derek C. Schuurman, "Technology and the Biblical Story," *Pro Rege* 46, no. 1 (September 2017): 4.

[6]Winston Churchill, address to House of Commons, October 28, 1943.

[7]John Culkin, "A Schoolman's Guide to Marshall McLuhan," *Saturday Review* 18 (March 1967): 51-53.

[8]"The lawnmower must not be used in particular to trim bushes, hedges and shrubs," John Deere, *JS63V Operator's Manual*, SAU11483, 2015, 4.

[9]Winston Ewert, "Is Technology Neutral? Or Does it Change Our World Whether We Like it or Not?," *Mind Matters*, October 30, 2018, https://mindmatters.ai/2018/10/is-technology-neutral/.

[10]3M, "History Timeline: Post-it Notes," https://www.post-it.com/3M/en_US/post-it/contact-us/about-us/.

[11]Edwin Black, *IBM and the Holocaust: The Strategic Alliance Between Nazi Germany and America's Most Powerful Corporation* (New York: Crown Publishers, 2001).

[12]Sarah Lewis, "The Racial Bias Built Into Photography," *New York Times*, April 25, 2019, https://www.nytimes.com/2019/04/25/lens/sarah-lewis-racial-bias-photography.html.

[13]William Robbins, "Engineers Are Held at Fault in '81 Hotel Disaster," *New York Times*, November 16, 1985, https://www.nytimes.com/1985/11/16/us/engineers-are-held-at-fault-in-81-hotel-disaster.html.

[14]Alan Yuhas, "Pokémon Go: Armed Robbers Use Mobile Game to Lure Players into Trap," *The Guardian*, July 11, 2016, https://www.theguardian.com/technology/2016/jul/10/pokemon-go-armed-robbers-dead-body.

[15]Marshall McLuhan, *Understanding Media: The Extensions of Man* (Cambridge, MA: The MIT Press, 1994), 91.

[16]Carl Mitcham, *Thinking Through Technology: The Path Between Engineering and Philosophy* (Chicago: University of Chicago Press, 1994), 77.

[17]Peter-Paul Verbeek, "Beyond Interaction: A Short Introduction to Mediation Theory," *Interactions*, May-June 2015, 26-31.

4 FAITHFUL DESIGN GUIDES

[1]Stephen V. Monsma, ed., *Responsible Technology* (Grand Rapids, MI: Eerdmans, 1986), 182-83.

[2]W3C, "Web Content Accessibility Guidelines," https://www.w3.org/TR/WCAG20/.

[3]Albert M. Wolters, *Creation Regained: Biblical Basics for a Reformational Worldview* (Grand Rapids, MI: Eerdmans, 1985), 28.

[4]Charles W. Colson and Nancy Pearcey, *The Christian in Today's Culture* (Wheaton, IL: Tyndale House, 1999), 317.

[5]Monsma, *Responsible Technology.*

[6]Egbert Schuurman, "Responsible Ethics for Global Technology," *Axiomathes* 20, no. 1 (March 2010): 107-27.

[7]Gayle E. Ermer and Steven H. VanderLeest, "Using Design Norms to Teach Engineering Ethics," *Proceedings of the 2002 American Society for Engineering Education Annual Conference*, Montreal, Canada, 2002.

[8]Derek C. Schuurman, *Shaping a Digital World: Faith, Culture and Computer Technology* (Downers Grove, IL: IVP Academic, 2013).

[9]T. S. Eliot, "Choruses from the Rock," *Collected Poems 1909-1962* (London: Faber & Faber, 2009).

[10]L. Kalsbeek, *Contours of a Christian Philosophy*, ed. Bernard and Josina Zylstra (Toronto, ON: Wedge Publishing Foundation, 1975), 192.

[11]Albert Borgmann, *Power Failure: Christianity in the Culture of Technology* (Grand Rapids, MI: Brazos), 2003.

[12]Franz Kafka, *The Trial*, trans. David Wyllie (Teddington, UK: Echo Library, 2007), 1.

[13]Keith Barry, "Making Cars Safer for Women," *Consumer Reports*, February 2020, 52-58.

[14]Barry, "Making Cars Safer for Women."

[15]Cathy O'Neil, *Weapons of Math Destruction* (New York: Broadway Books, 2017), 145-46.

5 BEYOND ENGINEERING ETHICS

[1]This case study was inspired by one reported in Stephen H. Unger, "Examples of Real World Engineering Ethics Problems," *Science and Engineering Ethics* 6, no. 3 (2000): 424-25. The name used in the story has been changed.

[2]C. S. Lewis, *Christian Reflections* (Grand Rapids, MI: Eerdmans, 2014), 41.

[3]National Society of Professional Engineers, "NSPE Code of Ethics for Engineers," https://www.nspe.org/resources/ethics/code-ethics.

[4]Engineers are often required to sign nondisclosure agreements (NDAs), but the scope of these agreements ought to be kept reasonable.

[5]The phrase "sanctified common sense" is used by Albert Wolters in *Creation Regained: Biblical Basics for a Reformational Worldview* (Grand Rapids, MI: Eerdmans, 2005), 36.

[6]These scenarios are variations of a classical ethical case study referred to as the "trolley problem."

[7]NSPE Code of Ethics, https://www.nspe.org/resources/ethics/code-ethics.

[8]N. T. Wright, *After You Believe: Why Christian Character Matters* (New York: Harper-Collins, 2010), 70.

[9]Steven H. VanderLeest, "Justice and Humility in Technology Design," *Proceedings of the 2006 American Society for Engineering Education (ASEE) Conference*, Chicago, IL, June 2006.

[10]Wright, *After You Believe*, 60.

[11]Wright, *After You Believe*, 70.

[12]Wright, *After You Believe*, 205.

[13]Wright, *After You Believe*, 70.

[14]Wright, *After You Believe*, 18-20.

[15]Wright, *After You Believe*, 21.

[16]Derek C. Schuurman, "Modern Devices and Ancient Disciplines," *Faith Today*, November/December 2017, 39-41.

[17]Quoted in Craig G. Bartholomew, *Contours of the Kuyperian Tradition: A Systematic Introduction* (Downers Grove, IL: IVP Academic, 2017), 317.

[18]For a discussion of some spiritual disciplines, see Rebecca Konyndyk DeYoung, *Glittering Vices: A New Look at the Seven Deadly Sins and Their Remedies*, 2nd ed. (Grand Rapids, MI: Baker, 2020), 220-31.

[19]C. S. Lewis, *The Abolition of Man* (New York: HarperOne, 1974), 18, 83-101.

[20]Lewis, *Abolition of Man*, 17.

[21]Alasdair MacIntyre, *After Virtue: A Study in Moral Theory*, 2nd ed. (Notre Dame, IN: University of Notre Dame Press, 1984), 216.

[22]N. T. Wright, "How Can the Bible Be Authoritative?," *Vox Evangelica* 21 (1991): 18.

[23]Bruce R. Ashford, *Every Square Inch: An Introduction to Cultural Engagement for Christians* (Bellingham, WA: Lexham Press, 2015), 19.

[24]For more discussion of norms within the context of engineering ethics, see Gayle E. Ermer and Steve VanderLeest, "Using Design Norms to Teach Engineering Ethics," *Proceedings of the 2002 American Society for Engineering Education*, 2002.

[25]Gayle Ermer, "Professional Engineering Ethics and Christian Values: Overlapping Magisteria," *Perspectives in Science and Christian Faith* 60, no. 1 (March 2008): 33.

[26]Gene Haas, "Kuyper's Legacy for Christian Ethics," *Calvin Theological Journal* 33 (Fall 1998): 320-49.

[27]Timothy Keller, *Every Good Endeavor: Connecting Your Work to God's Work* (New York: Penguin Books, 2016), 218.

[28]Langdon Winner, "Engineering Ethics and the Political Imagination," in *Broad and Narrow Interpretations of the Philosophy of Technology*, ed. Paul T. Durbin (Dordrecht, NL: Kluwer Academic Publishers, 1990), 53.

[29]Wright, *After You Believe*, 69.

6 MODERN TOWERS OF BABEL

[1]Daniel Allen Butler, *Unsinkable: The Full Story of the RMS Titanic* (Mechanicsburg, PA: Stackpole Books, 1998), 1.

[2]Butler, *Unsinkable,* 48.

[3]Butler, *Unsinkable,* 168.

[4]*Titanic,* Paramount Pictures, directed by James Cameron, released December 1997.

[5]Jacques Ellul, "Technique and the Opening Chapters of Genesis," in *Theology and Technology: Essays in Christian Analysis and Exegesis,* ed. Carl Mitcham and Jim Grote (Lanham, MD: University Press of America, 1984), 126.

[6]Jacques Ellul, *The Technological Society* (New York: Vintage, 1964), xxv.

[7]Craig Gay, *Modern Technology and the Human Future: A Christian Appraisal* (Downers Grove, IL: IVP Academic, 2018), 12.

[8]Neil Postman, *Technopoly: The Surrender of Culture to Technology* (New York: Vintage 1993), 52.

[9]Lewis Mumford, *Technics and Civilization* (Chicago: University of Chicago Press, 2010).

[10]Time and motion studies were part of scientific management theory, sometimes referred to as Taylorism after its founder Frederick W. Taylor. See Lee Hardy, *The Fabric of this World: Inquiries Into Calling, Career Choice, and the Design of Human Work* (Grand Rapids, MI: Eerdmans, 1990), 128-40.

[11]Lewis Mumford, *The Pentagon of Power: The Myth of the Machine* (New York: Harcourt Brace Jovanovich, 1970), 57.

[12]C. S. Lewis, *The Abolition of Man* (New York: HarperOne, 2015), 77.

[13]Derek Schuurman, "The Magic of Technology," *Christian Courier,* April 13, 2020, 16.

[14]Eugene Peterson, *As Kingfishers Catch Fire* (Colorado Springs, CO: Waterbrook, 2017), 95-96.

[15]Lynn White, "The Historical Roots of Our Ecological Crisis," *Science* 155, no. 3767 (March 10, 1967): 1203-1207.

[16]See https://www.rohsguide.com/; https://www.usgbc.org/leed; and https://www.aiche .org/efs/institute-sustainability.

[17]C. S. Lewis, *The Screwtape Letters* (New York: HarperOne, 2015), 135.

[18]Martin Buber, *I and Thou* (New York: Touchstone, 1996), 156.

[19]For a helpful description of idolatry, see Heidelberg Catechism Q/A 95.

[20]John Calvin, *Institutes of the Christian Religion,* vol. 1, ed. John T. McNeill, trans. Ford Lewis Battles (Philadelphia, PA: Westminster Press, 1960), 108.

[21]Timothy Keller, *Counterfeit Gods: The Empty Promises of Money, Sex, and Power, and the Only Hope That Matters* (New York: Riverhead Books, 2009), 64-65.

[22]G. K. Beale, *We Become What We Worship: A Biblical Theology of Idolatry* (Downers Grove, IL: IVP Academic, 2008), 11.

[23]C. S. Lewis, *The Voyage of the Dawn Treader* (London: Fontana Lions, 1980), 73.

[24]For examples of this kind of thinking, see Brian Christian and Tom Griffiths, *Algorithms to Live By: The Computer Science of Human Decisions* (New York: Henry Holt and Company, 2016).

[25]Neil Postman, *Technopoly: The Surrender of Culture to Technology* (New York: Vintage, 1992), 52.

[26]For a good discussion of this, see Peter-Paul Verbeek, *What Things Do: Philosophical Reflections on Technology, Agency, and Design* (University Park: Pennsylvania State University Press, 2005), 216-17.

[27]The notion of performing a "liturgical audit" is suggested by James K. A. Smith in *You Are What You Love* (Grand Rapids, MI: Brazos, 2016), 53-54, 114.

[28]Postman, *Technopoly*, 182.

[29]Ulf Lindqvist and Peter G. Neumann, "The Future of the Internet of Things," *Communications of the ACM* 60, no. 2 (February 2017): 26-30.

[30]Roger Forsgren, "The Architecture of Evil," *The New Atlantis* 36 (Summer 2012): 45.

[31]Forsgren, "The Architecture of Evil," 53.

[32]Albert Speer, *Inside the Third Reich* (New York: Touchstone, 1970), 375.

[33]Forsgren, "The Architecture of Evil," 44.

[34]Bob Goudzwaard, *Capitalism and Progress: A Diagnosis of Western Society* (Toronto, ON: Wedge Publishing Foundation, 1979), 243.

[35]See Matt Parker, *Humble Pi: When Math Goes Wrong in the Real World* (New York: Riverhead Books, 2020), 231-32.

[36]Dramatic footage of this event can be viewed online at https://archive.org/details/Tacoma-Narrows_Bridge_Collapse.

[37]Sara Baase, *A Gift of Fire: Social, Legal and Ethical Issues for Computing Technology*, 4th ed. (Upper Saddle River, NJ: Pearson, 2013), 377.

[38]For a detailed discussion of this, see Derek C. Schuurman, *Shaping a Digital World: Faith, Culture and Computer Technology* (Downers Grove, IL: IVP Academic, 2013), 67-70.

[39]Albert M. Wolters, *Creation Regained: Biblical Basics for a Reformational Worldview* (Grand Rapids, MI: Eerdmans, 2005), 59.

[40]Wolters, *Creation Regained*, 59.

[41]This example originally appeared in Derek C. Schuurman, "Technology and the Biblical Story," *Pro Rege* 46, no. 1 (September 2017): 6-7.

[42]Some of these descriptions and examples originally appeared in Derek Schuurman, "The Meaning of Technology," *Christian Courier*, January 12, 2015, 15.

7 MOVING FORWARD BY LOOKING BACK

[1]C. S. Lewis, *God in the Dock* (Grand Rapids, MI: Eerdmans, 1970), 200.

[2]David Kirsch, *The Electric Vehicle and the Burden of History* (Piscataway, NJ: Rutgers University Press, 2000), 31-32.

[3]Kirsch, *Electric Vehicle*, 17.

4The term *automobility* is used to describe the entire ethos, dream, or social ideology that surrounds the emergence of the twentieth-century American "car culture." See James Flink, *The Car Culture* (Cambridge, MA: The MIT Press, 1976).

5Robert A. Smith, *A Social History of the American Bicycle: Its Early Life and Times in America* (New York: American Heritage Press, 1972), 17-40.

6Alan Marcus, *Technology in America, A Brief History*, 3rd ed. (London: Palgrave, 2018), 143.

7Kirsch, *Electric Vehicle*, 129-66.

8David E. Nye, *Electrifying America: Social Meanings of a New Technology* (Cambridge, MA: The MIT Press, 1997), 96.

9Virginia Scharff, *Taking the Wheel: Women and the Coming of the Motor Age* (New York: Free Press, 1991), 36-37.

10Samuel B. Hays, "Gifford Pinchot and the American Conservation Movement," in *Technology in America: A History of Individuals and Ideas*, ed. Carrol Pursell, 3rd ed. (Cambridge, MA: The MIT Press, 2018), 155-65.

11Ruth Swartz Cowan, *A Social History of American Technology*, 2nd ed. (New York: Oxford University Press, 2018), 204.

12Kirsch, *Electric Vehicle*, 4.

13Samuel Florman, *The Existential Pleasures of Engineering*, 2nd ed. (New York: St. Martin's Griffon, 1994), 39.

14Florman, *Existential Pleasures of Engineering*, 39.

15Florman, *Existential Pleasures of Engineering*, 40.

16Aaron Clark, "Is This The Golden Age of Driving?," *Popular Science* 265, no. 3, September 2004, 97-103.

17Stephen V. Monsma, ed., *Responsible Technology* (Grand Rapids, MI: Eerdmans, 1986), 182-83.

8 TECHNOLOGY AND THE FUTURE

1For an amusing overview of some of these predictions, see Paul Milo, *Your Flying Car Awaits: Robot Butlers, Lunar Vacations, and Other Dead-Wrong Predictions of the Twentieth Century* (New York: HarperCollins, 2009).

2Sean F. Johnston, "Alvin Weinberg and the Promotion of the Technological Fix," *Technology and Culture* 59, no. 3 (July 2018): 620-51. One can observe the sudden rise in the usage of this phrase using the Google Ngram Viewer. See https://books.google.com/ngrams/graph?content="technological+fix".

3Rachel Carson, *Silent Spring* (Boston, MA: Houghton Mifflin Company, 1962).

4Mark Hamilton Lytle, *The Gentle Subversive: Rachel Carson, Silent Spring, and the Rise of the Environmental Movement* (New York: Oxford University Press, 2007), 189.

5Thomas S. Hischak and Mark A. Robinson, *The Disney Song Encyclopedia* (Lanham, MD: Scarecrow Press, 2009), 198.

[6]Judith Barad and Ed Robertson, *The Ethics of Star Trek* (New York: Harper Perennial, 2001), xiv.

[7]In *Star Trek*, Captain Kirk was the captain of the starship *Enterprise* and Dr. Spock was his first officer and science officer.

[8]This quote is taken from *Star Trek: The Original Series*, season 1, episode 28, "The City on the Edge of Forever," aired 1967.

[9]Byron Reese, *Infinite Progress: How the Internet and Technology Will End Ignorance, Disease, Poverty, Hunger, and War* (Austin, TX: Greenleaf Book Group, 2013).

[10]Some of the content from this section originally appeared in Derek C. Schuurman, "The Challenge of Transhumanism," *Proceedings of the Christian Engineering Conference*, Dordt University, 2019.

[11]Andy Crouch connects transhumanism and the bionic man in his article "The Bionic Man and the Body of Christ," *Christianity Today*, April 2019, 63-65.

[12]Jacob Shatzer, *Transhumanism and the Image of God: Today's Technology and the Future of Christian Discipleship* (Downers Grove, IL: IVP Academic, 2019), 40.

[13]Shatzer, *Transhumanism and the Image of God*, 56

[14]Shatzer, *Transhumanism and the Image of God*, 56.

[15]Francis Fukuyama, *Our Posthuman Future: Consequences of the Biotechnology Revolution* (New York: Picador, 2002), 208-10.

[16]Craig Gay, *Modern Technology and the Human Future: A Christian Appraisal* (Downers Grove, IL: IVP Academic, 2018), 205.

[17]Gay, *Modern Technology and the Human Future*, 205.

[18]Ray Kurzweil, *The Age of Spiritual Machines* (New York: Penguin, 1999), 142.

[19]Andy Crouch, *Playing God: Redeeming the Gift of Power* (Downers Grove, IL: InterVarsity Press, 2013), 64.

[20]Gay, *Modern Technology and the Human Future*, 131.

[21]Gay, *Modern Technology and the Human Future*, 178-79.

[22]Steven Garber, *Visions of Vocation: Common Grace for the Common Good* (Downers Grove, IL: InterVarsity Press, 2014), 59.

[23]Ronald Wright, *A Short History of Progress* (New York: Carroll & Graf, 2005), 108.

[24]Samuel C. Florman, *The Existential Pleasures of Engineering*, 2nd ed. (New York: St. Martin's Griffin, 1994), 87.

[25]Albert M. Wolters, *Creation Regained: Biblical Basics for a Reformational Worldview* (Grand Rapids, MI: Eerdmans, 2005), 61.

[26]Wolters, *Creation Regained*, 61.

[27]Richard Mouw, *When the Kings Come Marching In: Isaiah and the New Jerusalem* (Grand Rapids, MI: Eerdmans, 2002), 28.

[28]Lewis B. Smedes, *My God and I: A Spiritual Memoir* (Grand Rapids, MI: Eerdmans, 2003), 59.

9 MUST WE LEAVE OUR NEURAL NETS TO FOLLOW HIM?

[1]Walter M. Miller Jr., *A Canticle for Leibowitz* (Philadelphia: J. B. Lippincott & Co., 1959), 173.

[2]R. C. Sproul, *The Prayer of the Lord* (Lake Mary, FL: Reformation Trust, 2009), 51.

[3]US Equal Employment Opportunity Commission, "Questions and Answers: Religious Discrimination in the Workplace," https://www.eeoc.gov/policy/docs/qanda_religion.html.

[4]James Hatch, "Employee Fired For Religious Proselytizing," *Workforce*, December 1, 2006.

[5]Nicholas Wolterstorff, *Until Justice and Peace Embrace* (Grand Rapids, MI: Eerdmans,1983), 16.

[6]Bruce Ashford, *Every Square Inch: An Introduction to Cultural Engagement for Christians* (Bellingham, WA: Lexham Press, 2015), 5.

10 LETTERS TO A YOUNG ENGINEER

[1]Frederick P. Brooks, "The Computer Scientist as Toolsmith II," *Communications of the ACM* 39, no. 3 (March 1996): 68.

[2]These letters are the product of the author's imagination and no identification with actual persons, businesses, places, or locales should be inferred. This chapter was modeled after an article by Derek Schuurman titled "Letter to a Young Engineer" which appeared in Christian *Courier*, February 11, 2019.

[3]See https://dilbert.com/.

[4]Søren Kierkegaard, *The Sickness Unto Death: A Christian Psychological Exposition for Upbuilding and Awakening* (Princeton, NJ: Princeton University Press, 1983), 32-33.

[5]This was inspired by an article by Charlie Adams, "Interning in Babylon," Dordt ASME newsletter, February 2001.

[6]Lewis Smedes, *My God and I* (Grand Rapids, MI: Eerdmans, 2003), 59.

REFLECTION QUESTIONS

[1]Mike W. Martin and Roland Schinzinger, *Ethics in Engineering,* 4th ed. (New York: McGraw-Hill, 2004), 25.

IMAGE CREDITS

Figure 1.1. John T. Daniels / Wikimedia Commons

Figure 1.2. NASA / Wikimedia Commons

Figure 1.3. NASA / Wikimedia Commons

Figure 1.4. Google Patents, https://patents.google.com/patent/US1647A/en

Figure 3.1. Doug Kerr, Creative Commons license CC BY-SA 2.0 / Wikimedia Commons

Figure 3.2. Associated Press, July 19, 1981. AP File Photo 810719013. Used by permission

Figure 3.3. Sshay0100, Creative Commons license CC BY-SA 3.0 / Wikimedia Commons

Figure 4.1. Blake Patterson, Creative Commons license CC BY 2.0 / Wikimedia Commons

Figure 4.2. Julo / Blueshade / Wikimedia Commons. With inset image and editing by Jamin Ver Velde. Full image is available at https://cs.calvin.edu/activities/books/fieldguide/images/disk-brake.jpg

Figure 4.3. Library of Congress, 1940 November 7, LC-USZ62-46682

Figure 6.1. Engraving by Willy Stöwer, *Der Untergang der Titanic*, 1912 / Wikimedia Commons

Figure 7.1. 1912 Detroit Electric car ad / Wikimedia Commons

Figure 8.1. *Wall-E*, directed by Andrew Stanton (Emeryville, CA: Pixar Animation Studios, 2008), https://movies.disney.com/wall-e

Figure 9.1. Uberprutser, Creative Commons License CC BY-SA 3.0 NL / Wikimedia Commons

GENERAL INDEX

SCRIPTURE INDEX

Finding the Textbook You Need

The IVP Academic Textbook Selector
is an online tool for instantly finding the IVP books
suitable for over 250 courses across 24 disciplines.

ivpacademic.com